T0361245

# Converting Land from Rural to Urban Uses

This title aims to use social science research to contribute towards solving policy problems raised by the rural to urban land conversion process and by high land prices in particular. Ultimately, this book aims to develop the information useful to public decisions on zoning, taxation, public investments, transport systems, new towns, and so on, as they might affect the cost and quality of the conversion process. This book will be of interest to students of environmental studies.

# Converting Land from Rural to Urban Uses

A. Allan Schmid

RFF PRESS
RESOURCES FOR THE FUTURE

First published in 1968
by Resources for the Future, Inc.

This edition first published in 2015 by Routledge
2 Park Square, Milton Park, Abingdon, Oxon, OX14 4RN
and by Routledge
711 Third Avenue, New York, NY 10017

*Routledge is an imprint of the Taylor & Francis Group, an informa business*

**Publisher's Note**
The publisher has gone to great lengths to ensure the quality of this reprint but points out that some imperfections in the original copies may be apparent.

**Disclaimer**
The publisher has made every effort to trace copyright holders and welcomes correspondence from those they have been unable to contact.

A Library of Congress record exists under LC control number: 68016165


ISBN 13: 978-1-138-85749-0 (hbk)
ISBN 13: 978-1-315-71869-9 (ebk)

# CONVERTING LAND FROM RURAL TO URBAN USES

*by A. Allan Schmid*

RESOURCES FOR THE FUTURE, INC.
*1755 Massachusetts Avenue, N.W., Washington, D.C. 20036*

Distributed by
The Johns Hopkins Press, *Baltimore, Maryland 21218*

Resources for the Future is a non-profit corporation for research and education in
the development, conservation, and use of natural resources. It was established
in 1952 with the co-operation of the Ford Foundation and its activities since then
have been financed by grants from that Foundation.  Part of the work of Resources
for the Future is carried out by its resident staff, part supported by grants to
universities and other non-profit organizations. Unless otherwise stated,
interpretations and conclusions in RFF publications are those of the authors; the
organization takes responsibility for the selection of significant subjects for study,
the competence of the researchers, and their freedom of inquiry.

This book is one of RFF's land use and management studies which are directed
by Marion Clawson. The research was supported by a grant to Michigan State
University. A. Allan Schmid is associate professor of agricultural economics at
the University. The manuscript was edited by Hazel Eagle. The illustrations were
drawn by Frank and Clare Ford.

*Director of RFF publications,* Henry Jarrett; *editor,* Vera W. Dodds;
*associate editor,* Nora E. Roots.

# PREFACE

Urbanization is perhaps the dominant social, economic, and political movement in the contemporary American scene. Large numbers of people are moving to cities, and within cities to new locations; considerable acreages of land are converted from rural to urban uses; large capital investments have been made; substantial fortunes have been made through land speculation and development, and some have been lost. Fraud and bribery in public zoning and other actions have occurred or are widely believed to have occurred, especially in growing suburban areas. All in all, suburbanization is lively, exciting, and even engrossing—but not necessarily wholly desirable in all its ramifications. To one familiar with American land history all of this is strangely reminiscent of what took place during the nineteenth century. Not the least of these similarities is a general lack of understanding of the true nature of the processes involved, and a spate of poorly devised public measures to deal with the undesirable aspects of each.

During the nineteenth century westward expansion dominated U.S. politics, economics, and social structure. In 1800, the United States was a small group of thinly settled states along the Atlantic seaboard; the new nation had sovereignty over nearly all the territory westward to the Mississippi (except Florida) but did not really occupy much of it. The Louisiana Purchase, the purchase of Florida, the annexation of Texas, the accession of territory in the Southwest as a result of the war with Mexico, the treaty with Britain that established our claim to the Pacific Northwest, and the purchase of Alaska from Russia were the major steps in expansion of the nation to its present boundaries. The acquired lands were surveyed, property lines established, a system of land records put into operation, and a vast land disposal system inaugurated. More than 6 million patents conveyed more than a billion acres of land from public to private ownership in what was possibly the greatest real estate deal in world history—one which has left the expression "doing a Land Office business" imbedded in everyday American speech.

During this time a vast continent was settled, the forest largely cleared and burned, the prairies plowed, the local roads and the railroads built, and the towns and cities established. During the same period large numbers of immigrants came to our shores, at first predominantly from northern and western Europe but later from southern and eastern Europe and elsewhere, and were gradually absorbed into our economy, our society, and our political structure. In a century, more or less, an industrial nation came into being, though the greatest economic growth still lay ahead.

Throughout most of the nineteenth century the disposition of the public lands was a leading factor in national politics. How the lands were to be sold, on what terms, and in what size blocks were but some of the issues. The formation of states out of the federal territory was the subject of many a historic political struggle. In the mid-1830's the federal financial surplus, caused in large part by excessive speculation in public lands, was a major national issue.

There are two aspects of the westward expansion for which there are striking parallels today. First, settlement did not move smoothly across the country; rather, it moved somewhat fitfully, leaping across or around large areas. Scattered settlement was far more typical than solid or orderly settlement. Second, it was marked throughout by large-scale speculation in land. There seemed nothing incongruous to many political leaders of the day in speculating in land while simultaneously acting as legislators upon measures that would affect the value of such land. While land prices sometimes rose too slowly to provide a return commensurate with what could have been earned elsewhere, or sometimes did not rise at all, the land speculator was ubiquitous on the frontier.

There was a great volume of land law, and of spoken and written discussion of the subject. But there was comparatively little social science research, as we understand the term today, into the westward expansion and the settlement process. Today we know far more than did the contemporaries about the westward movement, about the land settlement and colonization which accompanied it, and about the numerous economic and social processes involved. We know that many of the land laws were ill-suited to the kinds of land to which they were applied or were ill-adapted to the times. For example, a homestead act was passed after most of the land to which it was best suited had already passed from public to private ownership; a timber culture act sought to promote afforestation in regions physically unsuited to it; unrealistic residence requirements were imposed in homestead and other acts; and homestead acreages were modified a generation or more later than would have been appropriate to the land then available.

Our interest in nineteenth century land history is rooted in concern for today's problems. Some of the lessons or experiences of the past have direct relevance to the suburban land development problems. Also, the results of past settlement processes are very much with us today and will be with us for a long time to come. Property lines established in the original cadastral surveys became road locations, and these have carried over into road and street lines; size of land holdings which developed under different land disposal laws have seriously affected suburban development in recent times; attitudes about land rights and land uses have carried over; and other illustrations could be cited. Land use at any date is very much a product of land history; adjustments to meet current conditions always must begin from a base inherited from the past.

Today we are in the midst of another great redistribution of the nation's population. The rural countryside is emptying. Over half of all counties lost population during the 1950's, many others lost rural population. These trends are still strong, and probably will continue through the seventies. It is true that the national totals still show rural population as about constant; but when these data are disaggregated, gains nearer the larger cities are shown to cover up extensive losses in most rural areas. At the same time, our larger urban population aggregations are mushrooming. Equally profound population redistribution is taking place within the urban concentrations. The older city center—the whole central city, in many metropolitan areas—has lost population. In the well-publicized flight to the suburbs, large areas of suburban settlement have been established since the war.

This postwar population redistribution has unfortunately involved a substantial segregation of population along racial, income, age, and class lines. The core of many larger cities has come to be occupied predominantly by Negroes and by single persons (usually quite young) and couples without children (usually comparatively old). The central cities, that is, are coming increasingly to include the very poor and the relatively rich. The suburbs are inhabited overwhelmingly by white, middle class, married couples with children. While the social amalgamation process of the nineteenth century absorbed the immigrant into a common social structure, the twentieth century settlement process is dividing the nation and sharpening race, class, and income contrasts. Its long-run consequences may be very serious indeed.

Urbanization has involved a land conversion process different from, but perhaps as striking as, the conversion of forest or prairie to farm. Farm, forest, and other rural land has been reshaped physically, and much has been covered with asphalt, roofs, or other impervious surfaces with consequent interference with the hydrologic cycle and other natural processes. Farm property lines have been materially altered, and land title transfers have been dominant. Electric power lines, sewer lines, water supply pipes, schools, and numerous other physical improvements have drastically affected the use of the land.

Scattered settlement typifies this new settlement process, which is often called "sprawl." Some tracts of land are shifted from rural use to a relatively intensive urban use, while neighboring and intermingled tracts are left unimproved or fully idle. Land speculation remains ubiquitous; land does not typically go from the ownership of an operating farmer into the ownership of the residential builder. Instead, it goes by diverse and sometimes devious routes through a series of ownerships. Land prices, as Mr. Schmid shows, are often bid up substantially in the process.

The postwar urbanization period has been marked by more research into the land settlement process than took place a century ago. A characteristic of our modern culture is its research—which, unfortunately, does not always seem to add to our wisdom. Extensive planning, zoning, subdivision control, and other public action has aimed at control of the direction, speed, and final character of this urbanization process. Generally speaking, such public efforts have not been notably successful—no more than were many of the land laws of the nineteenth century. The urban planner and urban economist of today are almost always totally ignorant of our agrarian past—and sometimes scornful of it. Although we are an urban nation today we were once a rural nation, and our rural heritage remains with us; urban specialists would do well to familiarize themselves with it.

We, who are contemporary to massive urbanization in the United States, understand the social, political, and legal aspects of current land settlement little better than did the contemporaries of the westward expansion. The suburban land market does not behave like the traditional model of the competitive market, nor even like the traditional models of farmland market. Some of the public actions, and even more some of the proposals, seem to be based upon a concept of the suburbanization process which does not adequately explain what is taking place.

The postwar suburbanization process will leave its mark many decades hence. The day will come when the houses built in the past 20 years must be torn down and replaced. The problems of those latter day developers, and of the people who will inhabit the new houses, will depend in considerable part upon the kind of surburbs we are developing today. One should judge a land colonization or development process not only in terms of its suitability for the conditions of the day, but also in terms of its probable adaptability to meet the changed conditions of tomorrow.

Support of Mr. Schmid's study is one reflection of Resources for the Future's desire to contribute to a better understanding of the suburban settlement process. Obviously, we too are contemporaries, and there is no magic entry into the perspectives of history. But we hope that our research efforts may be helpful. We do not seek to develop or to advocate a hard and fast program for influencing the direction of suburban growth in the United States. Rather, we seek to provide useful facts and interpretations to those who would like to understand better what is really happening, as well as to those who might like to change the direction and form of urban growth in the United States. While we hope that our research may be pronounced "good" by our contemporaries, and that it may be suggestive to them, we do not aspire to providing the last word on a dynamic process.

It is in this spirit that we offer A. Allan Schmid's study to the interested professional public. He presents some interesting data and some stimulating ideas which merit consideration. He would be the first to agree that this report does not exhaust the subject nor present a fully comprehensive treatment of it. When one pushes through the frontier, his first road is not a broad paved highway. But one must begin somewhere, and he has made a most promising beginning. It is hoped that other studies under way by RFF staff and through grant projects will supplement and extend his work.

September, 1967

MARION CLAWSON
Director of RFF Studies in
Land Use and Management

## ACKNOWLEDGMENTS

I am especially indebted to Marion Clawson of Resources for the Future for his counsel in the conduct of this study. Important data were acquired through the helpful cooperation of I. Lee Amann, Division of Research and Statistics, Federal Housing Administration, and N. H. Rogg, National Association of Home Builders.

Thanks go to the following who reviewed the manuscript at various stages: Mortimer Kaplan, Department of Housing and Urban Development; Donald Shoup, Institute of Public Administration; Mason Gaffney, Department of Economics, University of Wisconsin—Milwaukee; Shirley Weiss, Center for Urban and Regional Studies, University of North Carolina; Robert C. Colwell, Department of Housing and Urban Development; G. M. Neutze, Australian National University; Irving Hoch, Resources for the Future; David Allee, Department of Agricultural Economics, Cornell University; and Jerome Pickard, Urban Land Institute. I also want to acknowledge the assistance of L. L. Boger, Department of Agricultural Economics, Michigan State University.

A. Allan Schmid

# TABLE OF CONTENTS

## LIST OF TABLES

## LIST OF APPENDIX TABLES

## LIST OF FIGURES

# CONVERTING LAND FROM RURAL TO URBAN USES

# INTRODUCTION

Ours is an increasingly urban population, which demands that more and more land be converted from rural to urban use. This conversion process is having far-reaching effects that will enhance or detract from many lives for years to come. But it is also becoming increasingly expensive, and dissatisfaction with its results is beginning to be heard.

How can social science research contribute to solving policy problems raised by the rural to urban land conversion process in general, and by high land prices in particular? This monograph is directed to the question of how such research should be oriented in the future.

The problems must be outlined and points of policy leverage identified. The objectives are to indicate where data and research are needed, and to suggest relevant hypotheses, test some of these and indicate how others might be tested. Ultimately, the monograph looks toward development of information useful to public decisions on zoning, taxation, public investments, transport systems, new towns, and so on, as they might affect the cost and quality of the conversion process.

Of what significance are rapidly increasing new lot prices and rapidly increasing percentage appreciation over farm value? To begin with, they affect the chances of people becoming home owners and the quality of housing they can afford. A National Association of Home Builders (NAHB) survey of its builder members found that builders in 1964 considered lack of market their most important problem. The Association's news letter states:[1]

> A valid argument can be made that this land price increase is partly responsible for difficulties in selling new homes. It is the price of land which determines the sale price of units to be built. An average $4,567

[1] Michael Sumichrast, "Land Costs are Rising, Survey Confirms," *Economic News Notes,* Special Report, 65-5 (Washington: National Association of Home Builders, 1965).

lot would most likely mean a home selling for $22,500–$25,000. It is obvious that this house is too expensive for the large number of families found in the lower- and middle-income groups. So the increase in the cost of land is a factor in pricing much of the public out of the market.

Some idea of the effect of high land prices is suggested by the cost of the lot for even the most modestly-priced housing. Table 1 shows the median market price of the site for various levels of the total price of new housing (including site) with Federal Housing Administration (FHA) insured mortgages. In 1963 the median market price of the site was $2,048 for the total housing property value category of $8,000 to $9,000. The site represented 24 per cent of the total value.

Clawson et al. estimate that the acreage in urbanized areas will increase from the 16.6 million acres of 1950 to 30.3 million in 1980 and 41 million in 2000.[2] Not all of this acreage will be developed for residential lots, but the figures give a rough idea of the acreage directly affected by the price appreciation that reflects urban demands.

If the 25.4 million additional acres needed by the year 2000 were to be conservatively valued at the 1964 average raw land price per acre ($3,878) paid by housing developers, they would represent a value of $98,501 million.

Table 1.  Percentage Distribution of FHA-Insured New Homes by Total Property Value and Market Price of Site, 1963

| FHA estimate of total property value | Percentage distribution, 1963 | Median market price of site, 1963 | Site price as percentage of total value,[a] 1963 |
|---|---|---|---|
| (dollars) | (per cent) | (dollars) | (per cent) |
| Less than 8,000 | 0.4 | 2,117 | — |
| 8,000 to 8,999 | 0.7 | 2,048 | 24.1 |
| 9,000 to 9,999 | 1.6 | 1,937 | 20.4 |
| 10,000 to 10,999 | 3.2 | 2,043 | 19.4 |
| 11,000 to 11,999 | 4.7 | 1,960 | 17.0 |
| 12,000 to 12,999 | 7.7 | 2,220 | 17.8 |
| 13,000 to 13,999 | 10.1 | 2,391 | ·17.7 |
| 14,000 to 14,999 | 11.8 | 2,553 | 17.6 |
| 15,000 to 15,999 | 12.4 | 2,764 | 17.8 |
| 16,000 to 16,999 | 11.5 | 2,865 | 17.4 |
| 17,000 to 17,999 | 8.8 | 3,146 | 18.0 |
| 18,000 to 18,999 | 7.0 | 3,348 | 18.1 |
| 19,000 to 19,999 | 5.1 | 3,622 | 18.6 |
| 20,000 to 21,999 | 6.7 | 4,028 | 19.2 |
| 22,000 to 24,999 | 5.2 | 4,542 | 19.7 |
| 25,000 or more | 3.1 | n.a. | — |

n.a. = Not available.
[a] Median price of site/mid-point of total property value range X 100.
Source: AR63, Table 46, 3/27/64 and AR64, Table 41a, 3/18/65, Federal Housing Administration. National data for single family homes insured by the FHA.

[2] Marion Clawson, R. Burnell Held, and Charles H. Stoddard, Land for the Future (Baltimore: The Johns Hopkins Press for Resources for the Future, Inc., 1960), p. 110.

This figure, large as it is, comprises only a portion of the total value of finished lots at the end of the conversion process. If that total could be reduced even slightly by improvements in the land conversion process, a significant amount of future expenditure would be saved.

The first section of this monograph will be concerned with the prices of finished sites for residential single-family buildings. The second will examine the components of these finished lot prices—opportunity cost of the farm land used, and development costs such as layout, grading, and installation of improvements. Estimates will then be made of the appreciation in value of residential sites above agricultural value of the land and development costs for various sites in the United States. Another question that will be explored is what factors contribute to the appreciation, including rental values and the possibility of monopoly profits.

Another problem concerns the land conversion industry's products. Cost has meaning only with respect to a specified product. Changes in lot size and quality of improvement are conceptually easy to account for. But the suburban commodity produced has some negative characteristics, including sprawl, uncomplementary mixes of use, high cost utility services, costly transportation features, unsightly views, and other features of the physical environment. The last part of the monograph will discuss whether the mix of products now being offered is the one dictated by consumer tastes within the limits of consumer resources, or whether certain legitimate demands are frustrated by the structure of the suburban land market. Related questions are who gets the appreciation in land values noted above and what difference its distribution makes for the kind of suburban subdivisions the land conversion industry is producing.[3]

---

[3] For an earlier exploration of this whole area, see Marion Clawson, "Urban Sprawl and Speculation in Suburban Land," *Land Economics*, May 1962. Also see Mason Gaffney, "Urban Expansion—Will It Ever Stop?," *Yearbook of Agriculture* (Washington: U.S. Department of Agriculture, 1958).

# II

# PRICES AND APPRECIATION

## SUBURBAN LAND VALUES OVER TIME AND AMONG CITIES

Now let us turn to a detailed examination of the prices of sites for new homes, over time and among different cities, and then to an analysis of the components of the total value of these sites.

The FHA maintains a price series for the site value of developed lots used for new single-family houses with FHA insured mortgages.[1] This series advanced more than 300 per cent between 1946 and 1964, which contrasts with an only 58 per cent increase in the general price level. The FHA reports that, on the average, land accounted for about 11 per cent of the total value of a new house in 1946, and for about 19 per cent in 1964 (see Table 2).

The FHA site value data are also available for selected housing areas across the U.S. (Boundaries conform to Standard Metropolitan Statistical Areas [SMSA] when the area is so designated, otherwise to county boundaries.) A second series is available from the National Association of Home Builders. Both series are included in Appendix Tables A–1, A–2, and A–3.[2]

---

[1] These prices cover primarily houses built in newly-developed tracts in the suburban areas at the growing edge of the city. Estimating bias is minimal, since the prices are based on appraisals used in making important loan decisions. The sites concerned are generally regarded as representing the prices of improved residential building sites sold to medium-income purchasers. The percentage distribution of houses by various total prices is shown in Table 1. Because of inter-area variability in lot size and costs of street improvements and rough grading, the inter-city data have some comparability limitations, but represent the best data available.

[2] A new case study of land prices for Philadelphia makes data available on the change in price over the period 1945–1962 by distance zones, proximity to bus lines, public facilities, zoning, and owner characteristics. Unfortunately, the study was not available when this monograph was written. See Grace Milgram, *The City Expands* (Philadelphia: University of Pennsylvania Institute for Environmental Studies, 1967).

Table 2.   Prices of Sites for Single-Family Houses Insured by FHA and for Farm Land, and the Ratio of Site to Total Property Value, United States Average, 1946–1967

| Year | NAHB data[a] Market price of site New | FHA data[c] Market price of site[b] New | FHA data[c] Market price of site[b] Used | FHA data[c] Site per cent of total price of house and lot New | FHA data[c] Site per cent of total price of house and lot Used | USDA data[d] Average value of farm real estate per acre |
|---|---|---|---|---|---|---|
| | *(dollars)* | *(dollars)* | | *(per cent)* | | *(dollars)* |
| 1946 | n.a. | 761 | 833 | 11.5 | 13.3 | 54 |
| 1947 | n.a. | 893 | 915 | 11.8 | 12.7 | 60 |
| 1948 | n.a. | 1,049 | 970 | 11.7 | 12.0 | 65 |
| 1949 | n.a. | 1,018 | 1,098 | 12.0 | 12.1 | 67 |
| 1950 | n.a. | 1,035 | 1,150 | 12.0 | 12.4 | 65 |
| 1951 | n.a. | 1,092 | 1,222 | 12.1 | 12.0 | 75 |
| 1952 | n.a. | 1,227 | 1,296 | 12.0 | 12.4 | 82 |
| 1953 | n.a. | 1,291 | 1,461 | 12.7 | 12.9 | 83 |
| 1954 | n.a. | 1,456 | 1,591 | 13.4 | 13.3 | 82 |
| 1955 | n.a. | 1,626 | 1,707 | 13.5 | 14.3 | 85 |
| 1956 | n.a. | 1,887 | 1,931 | 14.2 | 15.2 | 90 |
| 1957 | n.a. | 2,148 | 2,041 | 14.9 | 15.7 | 97 |
| 1958 | n.a. | 2,223 | 2,150 | 15.5 | 16.5 | 103 |
| 1959 | n.a. | 2,372 | 2,357 | 16.2 | 17.9 | 111 |
| 1960 | 2,808 | 2,477 | 2,354 | 16.7 | 17.7 | 116 |
| 1961 | n.a. | 2,599 | 2,503 | 17.2 | 18.3 | 118 |
| 1962 | n.a. | 2,725 | 2,721 | 17.6 | 19.1 | 124 |
| 1963 | n.a. | 2,978 | 2,850 | 18.4 | 19.7 | 130 |
| 1964 | 4,567 | 3,130 | 2,981 | 18.9 | 20.2 | 137 |
| 1965 | n.a. | 3,442 | 3,218 | 20.0 | 20.9 | 146 |
| 1966 | n.a. | 3,627 | 3,254 | 20.2 | 21.0 | 157 |
| 1967 1st qtr. | n.a. | 3,725 | n.a. | n.a. | n.a. | n.a. |

n.a. = Not applicable.
[a] *The 1964 NAHB Builder Membership Survey* (Washington: National Association of Home Builders, 1964), p. 22.
[b] Estimated price for an equivalent site including street improvements or utilities, rough grading, terracing, and retaining walls, if any.
[c] *FHA Annual Reports*, Federal Housing Administration, Division of Research and Statistics, Table 35 (revised 4/2/65).
[d] *Farm Real Estate Market Developments* (Washington: U.S. Department of Agriculture, CD–69, CD–67, CD–64, and CD–31).

The price of a building lot represents the cost of the total conversion. Its components include the opportunity cost of the land, namely, its agricultural value, the size of the lot, and such improvement costs as grading, streets, and utility lines. Any further appreciation of land value will be either profit or rent.

If the increase in lot prices over time and the differences among cities can be explained by variations in agricultural values, the implication is that consumers are forced to pay high prices because farm opportunity costs are high. If the higher prices are due to lot size and to quality of improvements, there is no great problem—people are merely buying larger and better lots. If they are due to increases in the cost of improvement, research should

Table 3.  Index Numbers for Prices of Sites for Single-Family Houses, Farm Value,
Construction Costs, and Consumer Prices, 1946–1964

(1957–1959 = 100)

| Year | Price of site[a] | Farm value[a] | Residential construction cost[b] (Boeckh) | Consumer prices (all items)[c] |
|---|---|---|---|---|
| 1946 | 34 | 52 | 57 | 68 |
| 1947 | 40 | 58 | 70 | 78 |
| 1948 | 47 | 63 | 78 | 84 |
| 1949 | 45 | 64 | 76 | 83 |
| 1950 | 46 | 63 | 80 | 83 |
| 1951 | 49 | 72 | 87 | 91 |
| 1952 | 55 | 79 | 89 | 93 |
| 1953 | 57 | 79 | 90 | 93 |
| 1954 | 65 | 79 | 90 | 94 |
| 1955 | 72 | 82 | 92 | 93 |
| 1956 | 84 | 87 | 97 | 95 |
| 1957 | 96 | 93 | 98 | 98 |
| 1958 | 99 | 100 | 99 | 101 |
| 1959 | 106 | 106 | 103 | 102 |
| 1960 | 110 | 112 | 104 | 103 |
| 1961 | 116 | 113 | 105 | 104 |
| 1962 | 121 | 119 | 106 | 105 |
| 1963 | 132 | 125 | 109 | 107 |
| 1964 | 139 | 132 | 112 | 108 |

[a] From Table 2.
[b] *Business Statistics* (Washington: U.S. Department of Commerce, 1965), p. 52.
[c] *Ibid.*, p. 39.

focus on the efficiency of the firms involved in land development. But if they are due to land value appreciation, then attention should shift to the reasons for this appreciation and, if necessary, to the means of control.

The U.S. Department of Agriculture maintains a price series for farm land values per acre. This series advanced about 150 per cent between 1946 and 1964. (For data, see Tables 2 and 3.) While the farm land price rise is significant, it is small compared with the lot price increase of over 300 per cent. While that part of total lot price represented by the farm value of the land is quite easy to obtain from existing data sources, the cost of improvements and changes in lot size over time is much less available. A residential construction cost index is available, but covers building costs and not land improvement. This index increased 96 per cent between 1946 and 1964.

The best case study available was made by Sherman J. Maisel in the San Francisco Bay area. He found that the typical single family lot rose in price from $1,300 to $3,850 between 1950 and 1962.[3] He estimated that 28

[3] Sherman J. Maisel, "Land Costs for Single-Family Housing," in *California Housing Studies* (Berkeley: Center for Planning and Development Research, University of California, 1963).

per cent of the increase in value figured on the original lot size was due to changes in lot size, and 52 per cent was due to the increase in the value of raw land (see Table 4). The raw land value per acre (computed as a residual) increased 221 per cent, from $3,300 to $10,000. The average value of all farm land in California increased 165 per cent, from $154 per acre in 1950 to $408 in 1962.[4]

Change in land values emerges as the most important contributor to the increase in lot prices, and it is not accounted for by changes in farm land opportunity costs.

Perhaps some insight can be gained into the factors contributing to the increase in lot prices by a summary of several studies of the cross-sectional differences in FHA site values among U.S. cities and states at one point in time or in terms of percentage change over time.

Some research studies have attempted to explain the cross-sectional differences in FHA site values among U.S. cities and states. While these studies

Table 4.   Typical San Francisco Bay Area Lot Prices, Raw Land Values, and Development Costs, 1950, 1960, and 1962

| Year | Size of lot | FHA lot value | Cost to develop | Raw land value | Land value as a percentage of total lot value | Cost to develop per front foot | Raw land value per acre[a] |
|------|-------------|---------------|-----------------|----------------|-----------------------------------------------|--------------------------------|----------------------------|
|      | (sq. ft.)   | (dollars)     | (dollars)       | (dollars)      | (per cent)                                    | (dollars)                      | (dollars)                  |
| 1950 | 5,500       | 1,300         | 700             | 600            | 46                                            | 10                             | 3,300                      |
| 1960 | 6,500       | 3,500         | 1,535           | 1,965          | 56                                            | 20                             | 9,300                      |
| 1962 | 6,500       | 3,850         | 1,615           | 2,235          | 58                                            | 21                             | 10,600                     |

Cause of value change                                                                                       *Per cent*
1950–1960   For original lot size
    $  635  in cost and quality of development      29
    $1,100  in value of raw land (5,500 sq. ft.)      50
    $  465  from change in size of lot (includes land cost and development cost)      21

    $2,200  total change in value ($3,500–$1,300)      100

1950–1962   For original lot size
    $  720  in cost and quality of development      28
    $1,325  in value of raw land      52
    $  505  from change in size of lot      20

    $2,550  total change in value ($3,850–$1,300)      100

[a] Computed as a residual.
*Source:* Sherman J. Maisel, "Land Costs for Single-Family Housing," in *California Housing Studies* (Berkeley: Center for Planning and Development Research, University of California, 1963).

[4] *Farm Real Estate Market Developments* CD–64 (Washington: U.S. Department of Agriculture, August 1963), p. 37.

focus on differences in total prices and not differences in percentage appreciation over farm value, they do suggest some areas of inquiry that will be examined later in this monograph.

The studies, by Gottlieb,[5] Maisel,[6] and Mittelbach and Cottingham,[7] all found that differences in lot prices were significantly related to differences in buyers' incomes. This may suggest the hypothesis that land sellers possess some degree of market power and are in a position to take advantage of higher incomes and charge all that the traffic will bear. To test this it would, however, be necessary to find out whether cities with higher incomes do not have proportionately larger and higher-quality lots.

The three studies were not so unanimous on other variables, but two of them indicated the significance of population. One found that differences in lot prices were related to the change in population in the previous ten years,[8] and another found them related to differences in total population.[9] One study found that population density was a significant variable.[10] However, the distribution of houses and jobs within the SMSA was not significant.[11] These results are not conclusive, but do indicate that the growth and the patterns of growth of cities are worth looking at for their impact on land values.

One study considered farm land values as a variable that explained cross-sectional SMSA land price differences and was significant in a multiple regression analysis.[12] Farm land values within the SMSA are of course more highly correlated with lot prices than are the values of all land in a state.

Urban development has been a factor influencing farm real estate values across the nation. For example, an empirical study in California, using data for 1939, 1949, and 1954, found a significant relationship between farm real estate values and variations in population pressure measured by total county population.[13] Another case study, of northern Santa Clara County, California, shows that sale prices of farm land increased from four to six

---

[5] Manuel Gottlieb, "Influence on Value in Urban Land Markets U.S.A., 1956–1961," *Journal of Regional Science*, Vol. 6, No. 1 (1965).

[6] Sherman J. Maisel, "Price Movement of Building Sites in the United States—A Comparison among Metropolitan Areas," *Regional Science Association Papers*, Vol. XII (1964), pp. 47–60.

[7] Frank G. Mittelbach and Phoebe Cottingham, *Some Elements in Interregional Differences in Urban Land Values*, Reprint No. 31, Real Estate Research Program (Los Angeles: University of California, June 1964).

[8] Maisel, *op. cit.*

[9] Mittelbach and Cottingham, *op. cit.*

[10] Maisel, *op. cit.*

[11] *Ibid.*

[12] *Ibid.*

[13] Vernon W. Ruttan, "The Impact of Local Population Pressure on Farm Real Estate Values in California," *Land Economics* (May 1961), pp. 125–31. Also see Edgar S. Dunn, *The Location of Agricultural Production* (Gainesville, Florida: University of Florida Press, 1954).

times between 1952 and 1962.[14] A study of Sacramento County, California indicated that land prices rose about $180 per year per acre over a three-year period ending in 1961.[15] The beginning price level was under $2,000.

Though the proximity to markets is a factor contributing to the value of land for agricultural purposes, the main causal effect seems to be running from urban pressures to farm value. Much of this land in SMSA's is priced much higher than the capitalization of farm income would warrant and, in fact, much is not used for agriculture at all while it awaits urban development.[16]

A pattern does seem to emerge from the above analyses (case, cross-sectional, and time-series), though additional work may be justified. There is a large and growing residual land value contributing to high lot prices which is not explained by agricultural opportunity costs, lot size, improvement costs, or general inflation.

Attention now turns more specifically to the question of how large this land value appreciation above farm and improvement costs might be, and then to an explanation of this appreciation as a basis for possible public action to control it.

## APPRECIATION OF LOT PRICES ABOVE FARM
## LAND COSTS AND IMPROVEMENT COSTS

Data to estimate suburban land value appreciation above agricultural opportunity costs[17] and development costs are marked by their scarcity. Returning to Maisel's San Francisco case study, it may be noted that for a typical lot the residual value of the land after deducting development costs from lot prices was $10,600 per acre in 1962.[18] The average value per acre of all agricultural land in California in the same year was $408, and

[14] The 1952 data were obtained by Jack Lessinger, *A Theory of Rural Urban Transition*, preliminary draft, May 1, 1962. The 1962 data were obtained by Harris and Allee (see fn. 15, below).

| Area | 1952 | 1962 |
|------|------|------|
| I–1  | $ 4,300 | $ 19,000—21,000 |
| I–2  | 2,900 | 12,000—13,000 |
| I–3  | 2,600 | 10,000—15,000 |
| O    | 1,900 | 8,500—10,000 |
| C    | 1,600 | 7,500—10,000 |
| D    | 900 | 5,000— 6,000 |

[15] Curtis C. Harris, Jr. and David J. Allee, *Urbanization and Its Effects on Agriculture in Sacramento County, California, Part 2. Prices and Taxes of Agricultural Land* (Berkeley: Giannini Foundation, Research Report No. 270, December 1963).

[16] While actual and substantial agricultural use of much suburban land is questionable, it may, nevertheless, be classified as technically in agricultural use according to state and local laws; especially as relates to tax purposes.

[17] It should be recognized that the present agricultural value of the land may not be the future value and may not have the correct relationship to farm income. Such variance is not likely to affect significantly any analysis of sizeable urban land appreciation. For an historical analysis of these relationships see C. R. Chambers, "Farm Land Income and Farm Land Value," *American Economic Review* (1924), pp. 673–698.

[18] Maisel, "Land Costs for Single-Family Housing."

the average value of irrigated orchards, vineyards, and groves was $2,301.[19] The higher figure amounts to an appreciation of value above agricultural opportunity costs of $8,290, or 359 per cent. This suggests the possibilities for public policies to affect conversion costs and particularly those due to land appreciation.

A case study is also available for a single subdivision of 159 lots in Lisle, Illinois in 1961.[20] The case is included here because it is one of the few which has detailed cost data (see Table 5). The average value of farm land

Table 5.   Land and Development Costs for a Lisle, Illinois Subdivision, 1961

(Average size lot, 12,180 sq. ft.)

| | | |
|---|---|---|
| Land cost to developer | $2,000 | per acre |
| Interest, 2 years, 12 per cent | 240 | |
| Taxes, 2 years | 40 | |
| | $2,280 | per acre |
| Land cost (2.63 lots/acre)ᵃ          $  867   per lot | | |
| Improvement cost (bare field) $1,797 | | |
| Sanitary sewer | $ 3.02 | / front ft. |
| Water main | 2.27 | / front ft. |
| Storm sewer | 3.56 | / front ft. |
| Concrete curb | 2.00 | / front ft. |
| Concrete sidewalk | 2.50 | / front ft. |
| Paving (½ of frontage), 35 ft. total width | 6.77 | / front ft. |
| Other improvement cost $638 | | |
| Engineering and supervision | 10 | per cent of bare field cost |
| Interest, 6 months | 6 | per cent of bare field cost |
| Repair and replacements | 2 | per cent of bare field cost |
| Developer overhead and profit | 20 | per cent of bare field cost |
| Total improvements costs | $2,435 | |
| Selling cost | 546 | |
| Total cost | $3,848 | |
| Sales price | 5,460 | |
| Profit | 1,612 | |
| (Illinois average farm land value per acre, $306) | Per lot calculations | |
| Farm value per lot (2.63 lots per acre) | $  116 | |
| Improvement cost | 2,435 | |
| Selling cost | 546 | |
| Total cost | $3,097 | |
| Lot sales price | 5,460 | |
| Appreciation | 2,363 | |
| Percentage appreciation above all costs | 76 | per cent |
| Percentage appreciation above farm value | 2,037 | per cent |

ᵃ The developer realized only 2.63 lots per acre because of 35-foot streets and extra land taken for a major thoroughfare and state highway right of way.

Source: Percy E. Wagner, A Critical Analysis of a Developing Subdivision, presented at the National Convention of the American Institute of Real Estate Appraisers, Miami Beach, Florida, 1961. Illinois average farm land value from U.S. Department of Agriculture, Farm Real Estate Market Developments (Washington, CD–64).

[19] Farm Real Estate Market Developments (Washington: U.S. Department of Agriculture, CD–66), p.16.

[20] Percy E. Wagner, A Critical Analysis of a Developing Subdivision, presented at the National Convention of the American Institute of Real Estate Appraisers, Miami Beach, Florida, 1961.

in Illinois in 1961 was $306. The developer paid $2,000 per acre for the land. On a per lot basis, the total appreciation in residual value above farm land values after all improvement costs were deducted amounted to $2,363 on lots priced at $5,460. This appreciation is 2,037 per cent of the farm land value.[21]

Since this type of data on improvement costs and changes in the size of lots is not available over time for other cities, only rough approximations can be made. But they will be made in order to illustrate the possibilities, using FHA and NAHB lot price data.

Two important data items are required before the amount of appreciation contained in lot prices compared with production costs can be calculated. First, the number of lots per acre of raw land must be known so that the agricultural value of the raw land used for each lot can be calculated. Secondly, the average size of lot in major cities plus the cost of its improvement must be known. The dangers of the residual approach employed here should be noted. Errors in any of the components show up in the residuals and may distort them. Firm conclusions must await better data inputs.

*Farm Land Opportunity Costs*

To make this analysis, two sets of lot price data will be used, one from FHA and the other from NAHB. For the FHA prices, data are available on the average number of lots per acre in each of the regions covered by FHA insuring offices.[22] They are computed by dividing the number of acres in subdivisions by the number of lots obtained from them. The U.S. average for 1964 and 1965 was about 3 lots per acre (see Table 6). These data are sufficient to calculate raw land costs per lot sold. The state agricultural value per acre divided by the number of lots per acre (in each area

[21] For comparative purposes, appreciation will be expressed as a percentage of the base agricultural value. This study is primarily concerned with changes over the base (which differs among areas). Calculations of percentage appreciation make it possible to compare meaningfully, for example, two parcels which each have an absolute increase in value of $200 over land initially valued for farm purposes at $100 and $1,000 respectively. Interest lies in the fact that the $200 increase is a 200% change in the former case and quite significant, while it is only a small percentage change for the higher priced land. Absolute levels of appreciation, however, are included in the following tables for the reader interested in different questions. The land values used in this study are the average for the state in which the housing area is primarily located. Census data are available on the average value of farm land and building by counties which could be matched to SMSA's and housing areas. While one would expect these values to be higher than the state averages used throughout this study, the figure cannot be interpreted as the agricultural opportunity costs. While it reflects a locational advantage for agricultural products, it also has built into it considerable anticipation of eventual urban values and thus is unsatisfactory as a basis for computing the appreciation in value in the land conversion process.

[22] The lots per acre data reflect regional differences in the use of land per lot, since they are for all of the cities included in the area of jurisdiction of the various insuring offices, while the FHA lot prices are for each of the specific SMSA's.

jurisdiction of the insuring offices) gives the agricultural value of the land contained in the lot and its share of the roads. This is shown in Appendix Table A–1 for 1964.

The other set of price data comes from NAHB. It does not include number of lots per acre corresponding to the lots for which NAHB prices

Table 6.   Number of Lots Per Acre, Selected Cities, 1962, 1964, and 1965[a]

| City | July–December 1962 | July–December 1964 | January–June 1965 | July–September 1965 | Average 1964–1965 |
|---|---|---|---|---|---|
| *Zone I* | | | | | |
| Conn., Hartford | 1.8 | 1.9 | 1.8 | 1.9 | 1.9 |
| Del., Wilmington | 2.9 | 2.8 | 2.6 | 3.3 | 2.9 |
| D.C., Washington | 1.7 | 2.5 | 2.7 | 2.3 | 2.5 |
| Md., Baltimore | 2.0 | 3.1 | 5.0 | 4.1 | 4.1 |
| Mass., Boston | 2.1 | 1.6 | 1.8 | 2.1 | 1.8 |
| N.H., Manchester | 2.6 | 0.8 | 2.1 | 2.4 | 1.8 |
| N.J., Camden | 2.7 | 2.3 | 2.5 | 2.5 | 2.4 |
| N.J., Newark | 2.8 | 1.5 | 1.5 | 2.9 | 2.0 |
| N.Y., Albany | 3.2 | 2.6 | 2.6 | 2.5 | 2.6 |
| N.Y., Buffalo | 2.9 | 2.9 | 2.6 | 3.5 | 3.0 |
| N.Y., Jamaica | 3.1 | 3.3 | 3.2 | 2.8 | 3.1 |
| Pa., Philadelphia | 3.9 | 4.0 | 4.3 | 9.3 | 5.9 |
| Pa., Pittsburgh | 3.3 | 3.2 | 2.5 | 2.9 | 2.9 |
| R.I., Providence | 2.5 | 1.4 | 3.3 | n.a. | 2.3 |
| Vt., Burlington | n.a. | n.a. | n.a. | 2.5 | 2.5 |
| *Zone II* | | | | | |
| Ala., Birmingham | 1.9 | 2.5 | 2.3 | 2.2 | 2.3 |
| Fla., Jacksonville | 3.0 | 2.8 | 1.8 | 2.7 | 2.4 |
| Fla., Miami | 3.2 | 3.5 | 3.4 | 2.9 | 3.3 |
| Fla., Tampa | 3.2 | 3.3 | 3.4 | 3.4 | 3.4 |
| Ga., Atlanta | 2.5 | 2.4 | 2.4 | 2.2 | 2.3 |
| Ky., Louisville | 3.0 | 3.4 | 3.4 | 2.6 | 3.1 |
| Miss., Jackson | 2.8 | 2.5 | 2.7 | 2.6 | 2.6 |
| N.C., Greensboro | 2.2 | 2.0 | 1.6 | 2.1 | 1.9 |
| S.C., Columbia | 2.3 | 2.6 | 2.7 | 2.1 | 2.5 |
| Tenn., Knoxville | 1.8 | 2.4 | 1.8 | 1.9 | 2.0 |
| Tenn., Memphis | 2.5 | 2.4 | 2.4 | 3.6 | 2.8 |
| Va., Richmond | 3.1 | 2.7 | 2.8 | 2.7 | 2.7 |
| W.Va., Charleston | 4.2 | 1.6 | 2.7 | 2.8 | 2.4 |
| P.R., San Juan | 4.6 | 4.2 | 6.2 | 4.0 | 4.8 |
| *Zone III* | | | | | |
| Ill., Chicago | 3.8 | 3.7 | 3.6 | 3.9 | 3.7 |
| Ill., Springfield | 3.8 | 4.1 | 3.6 | 2.9 | 3.5 |
| Ind., Indianapolis | 3.2 | 3.0 | 3.3 | 2.0 | 2.8 |
| Iowa, Des Moines | 3.7 | 3.1 | 3.6 | 4.1 | 3.6 |
| Mich., Detroit | 4.6 | 3.9 | 3.5 | 4.1 | 3.8 |
| Mich., Grand Rapids | 2.7 | 2.7 | 2.8 | 2.2 | 2.6 |
| Minn., Minneapolis | 2.3 | 2.3 | 3.0 | 2.5 | 2.6 |
| Neb., Omaha | 3.1 | 3.1 | 2.6 | 3.2 | 3.0 |
| N. Dak., Fargo | 3.1 | 2.0 | 2.1 | 2.2 | 2.1 |
| Ohio, Cincinnati | 3.3 | 2.6 | 2.4 | 2.4 | 2.5 |
| Ohio, Cleveland | 2.8 | 3.9 | 2.5 | 2.7 | 3.0 |
| Ohio, Columbus | 3.4 | 3.0 | 3.4 | 3.4 | 3.3 |
| S.Dak., Sioux Falls | 2.9 | 3.2 | 3.4 | 2.7 | 3.1 |
| Wis., Milwaukee | 2.5 | 2.7 | 2.7 | 5.1 | 3.5 |

Table 6.    (Continued)

| City | July–December 1962 | July–December 1964 | January–June 1965 | July–September 1965 | Average 1964–1965 |
|---|---|---|---|---|---|
| *Zone IV* | | | | | |
| Ark., Little Rock | 2.8 | 3.1 | 2.5 | 2.9 | 2.8 |
| Colo., Denver | 3.1 | 2.8 | 2.7 | 3.5 | 3.0 |
| Kan., Topeka | 2.5 | 2.5 | 2.7 | 2.3 | 2.5 |
| La., New Orleans | 3.5 | 4.2 | 3.9 | 4.0 | 4.0 |
| La., Shreveport | 2.2 | 2.8 | 2.6 | 2.1 | 2.5 |
| Mo., Kansas City | 3.4 | 3.2 | 3.3 | 2.9 | 3.1 |
| Mo., St. Louis | 3.1 | 3.3 | 2.9 | 3.1 | 3.1 |
| N.M., Albuquerque | 2.8 | 0.9 | 3.1 | 3.1 | 2.4 |
| Okla., Oklahoma City | 3.3 | 3.3 | 3.4 | 3.5 | 3.4 |
| Okla., Tulsa | 3.2 | 3.0 | 3.5 | 3.8 | 3.4 |
| Tex., Dallas | 3.6 | 4.0 | 3.2 | 3.3 | 3.5 |
| Tex., Fort Worth | 3.6 | 3.3 | 2.7 | 2.8 | 2.9 |
| Tex., Houston | 3.2 | 3.5 | 3.9 | 4.0 | 3.8 |
| Tex., Lubbock | 3.4 | 3.1 | 3.5 | 3.3 | 3.3 |
| Tex., San Antonio | 3.4 | 3.2 | 3.2 | 3.6 | 3.3 |
| *Zone V* | | | | | |
| Alaska, Anchorage | n.a. | 2.5 | 1.4 | 3.6 | 2.5 |
| Ariz., Phoenix | 2.9 | 3.1 | 3.2 | 3.6 | 3.3 |
| Calif., Los Angeles | 3.5 | 3.9 | 3.4 | 4.1 | 3.8 |
| Calif., Sacramento | 2.8 | 3.0 | 3.3 | 3.9 | 3.4 |
| Calif., San Diego | 3.7 | 3.5 | 4.0 | 3.0 | 3.5 |
| Calif., San Francisco | 5.5 | 3.9 | 3.7 | 4.2 | 3.9 |
| Calif., Santa Ana | 3.4 | 4.6 | 3.9 | 3.1 | 3.9 |
| Hawaii, Honolulu | 4.2 | 4.6 | 3.4 | 5.0 | 4.3 |
| Idaho, Boise | 3.5 | 3.2 | 3.7 | 3.4 | 3.4 |
| Mont., Helena | 1.7 | 2.1 | 2.3 | 1.4 | 1.9 |
| Nev., Reno | 4.3 | 3.6 | 3.9 | 4.1 | 3.9 |
| Ore., Portland | 3.2 | 2.8 | 2.9 | 3.4 | 3.0 |
| Utah, Salt Lake City | 3.5 | 3.4 | 3.5 | 3.5 | 3.5 |
| Wash., Seattle | 3.5 | 3.4 | 3.3 | 1.8 | 2.8 |
| Wash., Spokane | 2.6 | 3.0 | 3.4 | n.a. | 3.2 |
| Wyo., Casper | 3.8 | 4.2 | 3.7 | 0.9 | 2.9 |

n.a. = Not available.
[a] Data are for new housing.
  *Source:* Federal Housing Administration. The cities listed are where FHA area offices are located, but the figures will include averages of other cities in each area.

are available. So a fixed number of lots per acre will be used for illustrative purposes for all cities and for both 1960 and 1964.

The number of 2.6 lots per acre is chosen somewhat arbitrarily, on the basis of a Lisle, Illinois study which indicates that, for a lot of 12,180 square feet (84 feet by 145 feet), allowances for streets and corners required 3,220 square feet per lot.[23] This is computed for a street of 35-foot width. For this particular subdivision, an additional area of 1,320 square feet per lot was required for a major thoroughfare and a state highway. This meant that a 12,180-square-foot lot required a total area of 16,720 square feet, and allowed about 2.6 lots per acre. This figure is judged to be a conservative estimate and will be used for illustrative purposes in computing the agricul-

[23]Wagner, *op.cit.*

tural values shown in Appendix Tables A–2 and A–3. It will be seen that it is similar to the 3 lots per acre average for all U.S. FHA insuring offices in 1964–1965 shown in Table 6. The smaller number of lots per acre in the NAHB data seems justified, since probably more custom building on larger lots is included.

*Improvement Costs*

The second step in this analysis is to compute improvement costs as related to lot size. The FHA regional data on lots per acre cannot be used to calculate average lot size, since no allowances are made for land used in streets and so on.

No other data are available on the average lot size for all major U.S. cities, but some area data and scattered examples are available. It does appear that lot size increased substantially after World War II, but that the rate of increase has slowed down since the mid-1950's. (See Figure 1.)[24] Tunnard and Pushkarev obtained data from a number of cities, as shown in Table 7. Their best estimate of average U.S. lot size is 14,600 square feet.

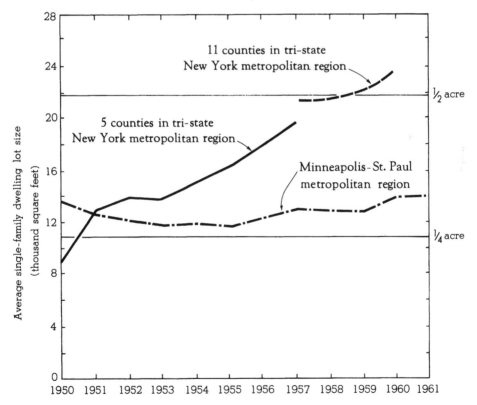

*Figure 1. Average lot size in subdivision approved, selected dates. (From Tunnard and Pushkarev; see footnote below.)*

[24] Christopher Tunnard and Boris Pushkarev, *Man-Made America* (New Haven: Yale University Press, 1963), p. 94.

Table 7. Lot Size in Selected Areas

| Place | Period covered | Lots recorded | Mean size (sq. ft.) | Percentage of lots | | |
|---|---|---|---|---|---|---|
| | | | | Small: below 9,999 | Medium: 10,000– 39,999 | Large: 40,000 & more |
| *Western areas:* | | | | | | |
| Los Angeles County (unincorp. territory) | 1959 | 25,982 | 14,000[a] | n.a. | n.a. | n.a. |
| Santa Barbara Co. | 1956–59 | n.a. | 22,000[a] | 60 | 13 | 26 |
| Pima Co., Ariz. (including Tucson) | 1954–59 | 28,278[b] | 15,560 | 83.9 | 11.7 | 4.4 |
| *Midwestern areas:* | | | | | | |
| Minn.-St. Paul Metro. area (6 counties) | 1950–58 | 71,024 | 12,300 | n.a. | n.a. | n.a. |
| Lansing, Mich. (city) | 1958–59 | n.a. | 16,000[a] | 50 | 45 | 5 |
| *Southern areas:* | | | | | | |
| Durham Co., N.C. (part only) | 1955–60 | 226[c] | 38,563 | 1 | 62 | 37 |
| *Middle Atlantic areas:* | | | | | | |
| Allegheny Co., Pa. | 1958–60 | 16,779 | 9,614 | 57 | 42 | 1 |
| Broome Co., N.Y. | 1950–58 | 8,599 | 11,500[a] | 75 | 20[a] | 5[a] |
| Delaware Co., Pa. | 1951–58 | 28,234[d] | 12,200[a] | 65.5 | 28.3 | 6.2 |
| Lebanon Co., Pa. | 1958–60 | 1,126 | 13,000[a] | 74 | 25[a] | 1[a] |
| *New York–New Jersey–Connecticut metropolitan region:* | | | | | | |
| 11 counties | 1958–60 | 36,778 | 22,120 | 35 | 56 | 9 |
| *New England towns:* | | | | | | |
| Canton, Mass. | 1959 | 244 | 20,100[a] | 13 | 85 | 1[a] |
| Guilford, Conn. | 1955–60 | 522[b] | 24,000[a] | 16 | 56.5 | 31.5 |
| West Hartford, Conn. | 1959–60 | 454 | 21,108 | 18.7 | 80.6 | .7 |
| Windsor, Conn. | 1954–60 | n.a. | 24,000[a] | 6 | 93 | 1 |

n.a. = Not available.
[a] Approximate.
[b] Building permits issued, rather than lots approved.
[c] Subdivided lots sold in sample area 25 square miles.
[d] Includes 34 per cent attached dwellings on lots smaller than 5,000 square feet.
*Source:* Christopher Tunnard and Boris Pushkarev, *Man-Made America* (New Haven: Yale University Press, 1963), p. 93.

They note, however, that the distribution of lot sizes within an area is highly skewed and, in the areas for which data were available, some 70 per cent of the newly-developed lots were less than 10,000 square feet.[25]

Even if data on the average lot sizes in all cities were available, there would be none on the cost of improving specific lot sizes.

However, certain studies are available. A detailed cost accounting of a new 159-lot subdivision in Lisle, Illinois in 1961 indicated that improvement costs for a 12,180 square foot lot (84 × 145) were $2,435.[26] This included sanitary and storm sewers, water main, paving, curb and gutter,

[25] *Ibid.*, p. 94.
[26] Wagner, *op. cit.*

sidewalk totaling $1,797, and also engineering, interest, and overhead of $638 (see Table 5). A survey of the Vancouver, British Columbia, region estimated that improvement costs for a 7,920 sq. ft. lot (66 × 120) were $2,000.[27] This included sewer, water main, paving, sidewalk, and street lights. A 1962 study of the San Francisco Bay area put the cost of a 10,000 square foot lot at $2,215.[28] See Table 8 for costs of developed lots of various sizes. And a 1955 study of the Boston area indicated development costs of $1,360 for a 10,000 square foot lot with an 80-foot frontage.[29] Costs do not include engineering, carrying costs, selling costs, and profit, which the study estimated might be 25 per cent of total development costs.

For illustrative purposes, a lot improvement cost of $2,435 will be used in all of the estimates of appreciation. This is the highest estimate indicated by the available studies.

*Lot Price Appreciation as a Percentage of Farm Value*

Computations of the appreciation of lot prices (less improvement costs) above agricultural value are shown in Appendix Table A–1 for 1964, using FHA data, and in Appendix Tables A–2 and A–3 for 1960 and 1964, using NAHB data. The computation process is as follows: for each city the analysis begins with the price per finished lot. The farm value of the land in the lot is computed and added to the lot improvement costs, and the total is subtracted from the finished lot price, to obtain the amount of absolute appreciation. This appreciation is then expressed as a percentage of the farm value. The FHA data show appreciation levels ranging from zero to 3,792 per cent of farm value for 1964, with an average of 892 per cent. The 1964 NAHB data range from zero[30] to 16,345 per cent, with an average of 1,875 per cent. It can be seen that if the estimates of improvement costs are anywhere near correct, appreciation as a per cent of farm value is very large in some cities. Comparison of the 1960 and 1964 NAHB data would suggest that appreciation percentages increased. Though it can only be guessed at with available data, the figures indicate that the improvement cost estimate was probably too high for 1960, since 100 out of the 259 cities studied did not have total lot prices covering estimated improvement costs plus farm costs, i.e., negative appreciation. However, in 1964 only 30 cities had total value less than these costs.

[27] *Land for Living* (New Westminster, B.C.: Lower Mainland Regional Planning Board of B.C., June 1963), p. 26.

[28] Maisel, *op. cit.,* p. 9.

[29] *The Effects of Large Lot Size on Residential Development* (Washington: Urban Land Institute, 1958). Front foot costs included roadway $4.00, water main $2.50, storm drainage $3.00, sanitary sewer $4.00, sidewalk $1.00, curbing $1.50, and loam, seeding, and street trees for curb strip $0.50.

[30] Some cities actually showed negative appreciations because of the high and constant estimate used for lot improvement costs. See following text for discussion of this.

Table 8. Cost of Developed Lots, San Francisco Bay Area, 1962

(Dollars)

| Square foot of lot[a] | Based on front feet[a] | Improvements | | Raw land per lot at $10,000 per acre | Total developed costs | Land at $10,000 per acre for 7,000 sq. ft. lots falling $600 per 1,000 sq. ft. | | Total costs per lot |
| | | Lot | Total per lot | | (3) + (4) | Per acre | Per lot | (3) + (7) |
| | (1) | (2) | (3) | (4) | (5) | (6) | (7) | (8) |
| 2,500 (row houses) | 525 | 175 | 700 | 835 | 1,535 | 12,700 | 1,060 | 1,760 |
| 5,000 | 1,050 | 200 | 1,250 | 1,670 | 2,920 | 11,200 | 1,760 | 3,010 |
| 6,000 | 1,260 | 225 | 1,485 | 2,000 | 3,485 | 10,600 | 2,120 | 3,605 |
| 7,000 | 1,470 | 350 | 1,720 | 2,330 | 4,050 | 10,000 | 2,330 | 4,050 |
| 8,000 at 75 front feet | 1,575 | 275 | 1,850 | 2,670 | 4,520 | 9,400 | 2,510 | 4,360 |
| 9,000 at 80 front feet | 1,680 | 300 | 1,890 | 2,860 | 4,840 | 8,800 | 2,520 | 4,410 |
| 10,000 at 90 front feet | 1,890 | 325 | 2,215 | 3,335 | 5,550 | 8,200 | 2,735 | 4,950 |

(2,500 through 7,000: 100 ft. depth)

*Source:* Sherman J. Maisel, "Land Costs for Single-family Housing," in *California Housing Studies* (Berkeley: Center for Planning and Development Research, University of California, 1963), p. 9.

[a] $21 per front ft. Includes street improvements or utilities and rough grading or terracing. 8,000 to 10,000 sq. ft. lots have increasing depths so that the frontages and costs of development do not rise as fast as size. Table assumes front foot costs account for corner lots waste, and so on, but net lots/acre decreased by 10 per cent to compensate for wasted space.

## APPRECIATION OF SUBURBAN RAW LAND PRICES
## ABOVE FARM LAND PRICES

The availability of lot improvement cost data is the weak link in the above analysis of appreciation computed as a residual from finished lot prices. The problem can be reduced by separating land appreciation into its two chronological parts, pre-development and post-development. One would expect that the appreciation contained in raw land prices paid by developers before any improvements are made would be less than the estimates made above for the total process. Nevertheless, the part of the total conversion process represented at the time the land is bought by an active developer who will build in a short time should provide a reliability check on the whole process.

### Raw Land Prices Paid by Developers

Fortunately, the National Association of Home Builders has conducted a survey of its member builders and has data on prices paid for raw land for residential building purposes in nine regions of the U.S. for 1960 and 1964 (see Table 9) and for 259 specific cities (see Appendix Tables A–7 and A–8).[31] The differences between these prices and farm values per acre give a rough measure of the appreciation over agricultural opportunity costs. This is shown in Appendix Tables A–7 and A–8. For quick reference, a few cities are also shown below in Table 10 to indicate the variability for 1964. The appreciation of value represented by the difference between the average values of urban and farm land acreages, expressed as a percentage of the average value of farm land, was 1,466 per cent in 1960 and 1,819 per cent

Table 9.  Prices Paid Per Acre of Raw Usable Land and Finished Lots,
U.S. Regions, 1960 and 1964

(*Dollars*)

| Region | Avg. price per acre 1960 | Avg. price per acre 1964 | Avg. price per lot 1960 | Avg. price per lot 1964 |
|---|---|---|---|---|
| Total U.S. | 2,447 | 3,878 | 2,808 | 4,567 |
| New England | 1,395 | 2,458 | 2,834 | 4,724 |
| Middle Atlantic | 3,077 | 4,966 | 3,219 | 5,148 |
| South Atlantic | 2,118 | 3,211 | 2,412 | 4,071 |
| East–South–Central | 1,533 | 2,420 | 2,250 | 3,564 |
| West–South–Central | 1,910 | 3,428 | 2,256 | 3,645 |
| East–North–Central | 1,957 | 2,833 | 3,008 | 4,551 |
| West–North–Central | 1,914 | 2,282 | 2,749 | 4,182 |
| Mountain | 2,309 | 3,552 | 2,404 | 3,798 |
| Pacific | 4,669 | 8,162 | 3,656 | 6,661 |

*Source:* National Association of Home Builders.

31 Data were collected from 7,100 respondents. In some small cities, however, the number of builders who answered was small and some untypical results may be represented. While these statistics represent the best data now available, this limitation should be kept in mind.

Table 10.   Suburban Raw Land Price Appreciation above Farm Land Prices,
Selected U.S. Cities, 1964[a]

| City and state | Suburban price/acre | Farm price/acre | Appreciation (1) − (2) | Appreciation over farm value (3) ÷ (2) |
|---|---|---|---|---|
| | (1) | (2) | (3) | (4) |
| Atlanta, Ga. | $ 1,791 | $127 | $ 1,664 | 1,310% |
| Berkeley, Cal. | 10,614 | 460 | 10,154 | 2,207 |
| Charlotte, N.C. | 2,633 | 234 | 2,399 | 1,025 |
| Cincinnati, Ohio | 3,696 | 282 | 3,412 | 1,211 |
| Cheyenne, Wyo. | 3,000 | 26 | 2,974 | 11,438 |
| Camden, N.J. | 3,820 | 600 | 3,220 | 537 |
| Dallas, Tex. | 7,277 | 108 | 7,169 | 6,638 |
| Dayton, Ohio | 2,780 | 282 | 2,498 | 886 |
| Des Moines, Iowa | 1,785 | 265 | 1,520 | 574 |
| Harrisburg, Pa. | 2,255 | 222 | 2,033 | 916 |
| Indianapolis, Ind. | 2,852 | 293 | 2,559 | 873 |
| Memphis, Tenn. | 3,681 | 165 | 3,516 | 2,131 |
| Minneapolis, Minn. | 2,160 | 168 | 1,992 | 1,186 |
| Omaha, Nebr. | 2,821 | 104 | 2,717 | 2,613 |
| Orlando, Fla. | 3,089 | 307 | 2,782 | 906 |
| Portland, Oreg. | 4,078 | 99 | 3,979 | 4,019 |
| St. Louis, Mo. | 4,673 | 139 | 4,534 | 3,262 |
| Syracuse, N.Y. | 2,600 | 165 | 2,435 | 1,476 |
| Springfield, Mass. | 1,731 | 349 | 1,382 | 396 |
| Washington, D.C. (Md.)[b] | 5,785 | 365 | 5,420 | 1,485 |
| Wichita, Kans. | 1,678 | 114 | 1,564 | 1,372 |

[a] For other cities and data sources, see Appendix Table A–8.

[b] The suburban area of Washington, D.C. includes both Maryland and Virginia. Maryland farm land price data were used because they reflected the highest value and would produce a conservative estimate of appreciation.

in 1964.[32] This compares with a 1960 and 1964 total appreciation of 399 per cent and 1,875 per cent in the total conversion process estimated on NAHB lot prices as shown in Appendix Tables A–2 and A–3. The 1964 total conversion process appreciation is larger than the partial appreciation (up to the development stage), as would be expected. However, the 1960 total appreciation is less than the partial appreciation. This suggests that the improvement cost estimates used in estimating the total conversion appreciation were too high for 1960. It should be noted that lack of data required that the same improvement cost level be used in both 1960 and 1964.

## Raw Land Prices Received by Farmers

Regional data on prices paid farmers for land that was to be used for nonfarm purposes is available from the USDA (see Table 11). Note that this is the price paid farmers for farm land, and that the land may be held

[32] In absolute terms, raw land prices increased about 50 per cent between 1960 and 1964. The average price was $1,995 in 1960 and $3,030 in 1964.

Table 11.  Market Value Per Acre of Land for Specified Nonfarm Uses, Selected Type of Farming Areas, October 1, 1961,[a] and for Farm Uses, March 1961

(Dollars)

| Type of farming area | Subdivisions Most frequent price[b] | Subdivisions Range[c] | Rural residences[d] Most frequent price[b] | Rural residences[d] Range[c] | Commercial, industrial Most frequent price[b] | Commercial, industrial Range[c] | Farm real estate |
|---|---|---|---|---|---|---|---|
| Northeast | 1,050 | 700–1,850 | 1,400 | 900–2,000 | 3,500 | 1,350–6,850 ⎫ | 190 |
| Eastern dairy | 2,000 | 1,900–2,400 | 1,200 | 950–1,600 | 3,350 | 2,700–4,200 ⎬ | 155 |
| Lake States dairy | 750 | 550–1,000 | 950 | 600–1,200 | 1,300 | 1,000–1,900 | 147 |
| General farming | 950 | 650–1,450 | 1,000 | 600–1,250 | 2,000 | 1,350–2,800 | 228 |
| Eastern corn belt | 1,400 | 900–1,850 | 1,250 | 1,250–1,750 | 2,850 | 2,450–6,500 ⎫ |  |
| Western corn belt | 1,050 | 850–1,350 | 1,100 | 850–1,350 | 1,650 | 1,300–2,400 ⎬ | 76 |
| Winter wheat | 1,050 | 800–1,250 | 1,200 | 550–1,050 | 1,150 | 800–1,350 | 139 |
| Eastern cotton | 700 | 500–950 | 650 | 500–700 | 3,550 | 2,000–4,800 | 126 |
| Central cotton | 850 | 550–1,350 | 650 | 450–1,000 | 2,050 | 1,300–3,500 | 89 |
| Western cotton | 1,300 | 1,250–1,700 | 1,100 | 950–1,100 | 2,150 | 1,500–2,900 | 43 |
| Northern range livestock | 1,300 | 950–1,750 | 1,200 | 900–1,750 | 4,000 | 2,600–5,750 | 244 |
| Northwest dairy | 750 | 550–1,100 | 1,050 | 850–1,400 | 2,500 | 1,400–3,000 | 234 |
| California specialty | 3,850 | 3,100–4,950 | 3,300 | 2,250–4,600 | 9,500 | 6,600–13,450 ⎫ |  |
| Florida | 1,650 | 1,050–2,800 | 1,200 | 750–1,550 | 5,200 | 2,400–12,700 ⎬ |  |
| Simple average[e] | 1,332 | | | | | | |

[a] Based on estimates supplied by farm real estate dealers and brokers for acreage tracts (not city lots) for the various uses specified.
[b] Average of "most frequent" prices reported, rounded to nearest $50.
[c] Averages of low and high ranges reported, rounded to nearest $50.

[d] Tracts of one acre or more for construction of a rural residence, not in a subdivision.
[e] Simple average of all regions (no weighting as between regions).
Source: Farm Real Estate Market Developments (Washington: U.S. Department of Agriculture, CD–60, March 1962), p. 20, and (CD–64, August 1963), p. 36.

various lengths of time before actual development. The NAHB data, on the other hand, show prices paid by developers, and some land may have been owned not by farmers but by various intermediate owners. The NAHB data may therefore be expected to show higher prices per acre than the USDA data.

## SUMMARY OF APPRECIATION LEVEL ANALYSIS

The various estimates of appreciation for 1960 and 1964 for both lot prices and raw land using FHA and NAHB data are summarized in Table 12. The table also shows the range of variation of individual cities within the overall averages. In comparing the two data sources it should be noted again that the FHA data generally represent medium priced housing while the NAHB data represent somewhat higher priced housing. The FHA data for 1964 show an 892 per cent appreciation in land value above farm prices (less improvement costs) while the NAHB data show 1,875 per cent. While these estimates are subject to data deficiencies in the components of lot size and improvement costs, the estimates of raw suburban land appreciation are not. The NAHB raw land data for 1964 show an appreciation of 1,819 per cent above farm values. From these data it can be concluded that appreciation levels are large and significant. Even if the farm land values were doubled or tripled, sizeable appreciation would remain.

The conclusion seems sound because it is drawn from the raw land price data alone, thus avoiding the data problems currently inherent in the analysis of appreciation through the whole conversion process to finished lots, though one would expect the appreciation in that case to be even larger. Though the quantification of the exact appreciation in the total

Table 12.  Summary of Percentage Net Appreciation of Finished Lots and Raw Land Prices above Farm Land Costs, 1960 and 1964

| Year | FHA data | | | NAHB data | | | |
|---|---|---|---|---|---|---|---|
| | | Percentage appreciation | | | Percentage appreciation | | |
| | Price | Range | Average | Price | Range | Average | Standard deviation |
| Finished lot prices (per lot) | | | | | | | |
| 1960 | $2,477 | n.a. | n.a. | $2,857 | 0–7,946[a] | 399 | n.a. |
| 1964 | 3,130 | 0–3,792[a] | 892 | 3,874 | 0–16,345[a] | 1,875 | n.a. |
| Raw land prices (per acre) | | | | | | | |
| 1960 | n.a. | n.a. | n.a. | 1,995 | 75–14,392 | 1,466 | 1,821 |
| 1964 | n.a. | n.a. | n.a. | 3,030 | 101–18,194 | 1,819 | n.a. |

n.a. = Not available.
[a] Some cities actually showed negative appreciations because of the high and constant estimate used for lot improvement costs. For discussion, see text.

conversion process awaits better data, it is clear that we are speaking of something which is often measured in hundreds and thousands per cent above farm land opportunity costs. Appreciation of this magnitude seems to deserve public policy attention.

When this appreciation is broken down in an attempt to show just how much is captured at each stage in the conversion process, the data are weak. Nevertheless, a summary table and chart are included here as a tentative first approximation (see Table 13 and Figure 2).

No claim is made that the tabulation is a valid statistical summary. However, it is the author's judgment that it does not represent an extreme case. For an acre of farm land valued at $300, the farmer might receive about $1,332. The active developer-builder might pay about $3,030, and add $6,331 in improvement costs. If the finished lots sold at about $10,072 per acre, the appreciation would be $3,441 per acre.

Now that significant levels of appreciation have been identified, analysis may proceed to the components of this appreciation and to factors associated with variation in appreciation levels among cities.

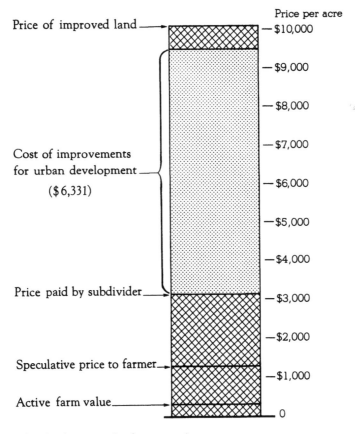

Figure 2. Land-price stages in the conversion process.

Table 13.   Land Prices at Various Stages in the Conversion Process: A Composite

|  | *(Dollars per acre)* |
|---|---:|
| Farm land value (1964)[a] | 300 |
| Price farmers received for subdivision use (1961)[b] | 1,332 |
| Price paid by developers for raw land (1964)[c] | 3,030 |
| Improvement cost ($2,435 x 2.6 lots/acre)[d] | 6,331 |
| Selling price of improved lots (1964) ($3,874 x 2.6 lots/acre)[e] | 10,072 |
| Total appreciation above farm land value (less improvement costs) | 3,441 |
| Percentage appreciation above farm land value[f] | 1,147% |

[a] A purposely high judgment of average U.S. farm land value weighted to those states with the most populous cities. The 1964 average value of farm land in the 48 states was $137.

[b] USDA data from Table 11. Simple average of regional averages without weighting.

[c] NAHB data from Appendix Table A–8.

[d] Data from Table 5.

[e] NAHB data from Appendix Table A–3.

[f] This composite produces a lower estimate of appreciation than the average shown in Appendix Table A–3 because of the higher farm land values used here.

# III

# COSTS OF CONVERSION—EFFICIENCY, RENTS, AND PROFITS

## CONVERSION EFFICIENCY

The very scarcity of production or land improvement cost data is indicative of the lack of research on the most efficient choice of inputs, materials, and practices. The scarcity contrasts with the wealth of production cost and efficiency studies for agricultural products.

Subdivision layout should also be studied because it affects the amount of land, utility lines, sewers, streets, and so on that will be needed, as should scale economies for these various improvement items. It would be useful to know how costs of production vary with the size of the contractors who construct sewers, paving, sidewalks, and other items, as well as what the landowner's other costs are. A survey of NAHB members indicates that the typical subdivision had 192 lots in 1964, but the figure varied from 300 lots in the mountain states to 115 lots in the New England states.[1]

Other organizational and institutional factors may affect improvement costs. One is whether the land development is done independently of the house building. The typical procedure prior to World War II is described by Clawson *et al.* as follows: "The process was subdivision by a 'developer' or speculator who sold, or tried to sell, the lots to individual buyers, each of whom would arrange to build as he chose and could finance."[2] This contrasts with present practices, where about 75 per cent of all builders build on their own land, which they purchase as improved lots or develop themselves.[3]

---

[1] *The 1964 NAHB Builder Membership Survey* (Washington: National Association of Home Builders, 1964), p. 3.

[2] Marion Clawson, R. Burnell Held, and Charles H. Stoddard, *Land for the Future* (Baltimore: The Johns Hopkins Press for Resources for the Future, Inc., 1960), p. 70. See also Philip H. Cornick, *Premature Subdivision and Its Consequences* (New York: Columbia University, 1938).

[3] NAHB, *op. cit.*, p. 8.

The cost of credit and risk to a firm which sells improved lots to builders rather than consumers, or which not only develops land but also builds and merchandises a total housing package, may be different from what it would be if the building and developing functions were separated. Other aspects of the risk element will be explored later.

Another way that the structure of the industry may affect costs is related to the firm's cash position. While eventual profits may be high, developers may be in a precarious cash position when sales commissions, promotional expenses, interest payments, and legal fees are high and must be paid currently.[4] Any organizational change which could reduce these costs and improve developers' cash flow position would be valuable.

The ability of financial institutions to make loan funds available for land purchase and development has often been limited by law or in practice. Institutional forms such as the land syndicate need to be explored for their implications for credit supply and costs.[5] The federal government has long guaranteed loans for home mortgages. The Federal Housing Administration was authorized to guarantee loans for private land development for the first time in 1965.[6] There has also been discussion of such loans for land purchase and development by state agencies.[7] Only in recent years has the Federal Home Loan Bank Board permitted member savings and loan associations to make land development loans.

## CONVERSION AND RENT

Some of the reasons for urban fringe land prices above farm value and improvement costs may now be explored. The buyer of a suburban lot acquires access to a bundle of valuable amenities, including locational advantage, air quality, association with neighbors, landscape features, and schools, playgrounds, and other public services. Some of these are limited in supply as a natural phenomenon, while others, though reproducible, are nevertheless limited in supply as a result of decisions not related to the cost of production. Returns above cost of production due to natural limitations in supply are technically termed *rents*, while such returns due to non-natural limitations (such as monopolistic restriction of new producers) are

[4] For an illustration, see Edward Elias, Jr., and William D. Warren, *The Unanchored Subdivision*, Real Estate Research Program, University of California, Los Angeles, December 1962 (mimeo). Summary published as *The Remote Subdivision*, Pamphlet No. 2, 1963.

[5] C. B. Singleton and W. H. Scofield, "Land Syndication and the Rural Urban Fringe," *Farm Real Estate Market Developments* (Washington: U.S. Department of Agriculture, 1962), CD-60.

[6] Housing and Urban Development Act of 1965, Public Law 89–117, 89th Cong., 1st sess., August 10, 1965, added the following title to the National Housing Act, "Title X—Mortgage Insurance for Land Development."

[7] Robert C. Weaver, Statement Before the Subcommittee on Housing of the Senate Committee on Banking and Currency on Housing and Urban Development Legislation, Washington, March 29, 1965.

termed *excess profits*. The latter should be distinguished from what are usually called *normal profits*, which include those payments necessary to draw forth the required entrepreneurial and capital resources.

This section will discuss such rent factors as locational advantage and landscape features. The next section is devoted to possible monopoly-like factors which could create excess profits.

Students of urban land values have long noted their relationship to distance from the central business district (CBD).[8] Users of close-in locations save transportation costs, and this advantage is capitalized into land prices. From this observation the expectation has arisen that in a static economy land values decline as distance from the CBD increases, until the land is no longer used for urban purposes, and its value is determined by agricultural opportunities. A theoretical rent or value gradient might look like that shown as line *A-B* in Figure 3. The rent gradient is related to the existing transportation system.

The point where existing houses or current construction stops may be called the rural-urban fringe or the marginal fringe site. (In practice this will not be a point or line but an area of some breadth.) In a static situation the marginal fringe site would also be the no-rent site for urban use, since it would offer no travel-saving advantage over any other urban site. However, it might receive agricultural rent.

In a dynamic situation, land beyond the marginal fringe site is expected to be developed in the future. This means that the current fringe site will in that future have a travel-saving locational advantage, and its present value is equal to its discounted future value, which exceeds the agricultural value. A land value gradient reflecting future developments might look like line *C-D* in Figure 3. Here the marginal fringe site *(B)* is no longer the no-rent margin. The no-rent margin has moved further out in the country to point *D*.

Part of the appreciation of fringe land values over farm value and improvement cost, according to this theory, might be explained as the present value of expected future land prices (reflecting future rents). The

[8] Lowdon Wingo, *Transportation and Urban Land* (Washington: Resources for the Future, 1961); Bernard Frieden, "Local Preferences in the Housing Market," *Journal of the American Institute of Planners*, November 1961, pp. 316–24; W. B. Hansen, "An Approach to the Analysis of Metropolitan Residential Extension," *Journal of Regional Science*, Summer 1961, pp. 37–55; Herbert Mohring, "Land Values and the Measurement of Highway Benefits," *Journal of Political Economy*, June 1961, pp. 236–49; William Alonso, "A Theory of the Urban Land Market," *Papers and Proceedings of the Regional Science Association*, 1960, pp. 149–57; and Eugene F. Brigham, "The Determinants of Residential Land Values," *Land Economics*, November 1965, pp. 325–35; Warren R. Seyfried, "The Centrality of Urban Land Values," *Land Economics*, August 1963, pp. 275–84; Richard F. Muth, "Economic Change and Rural-Urban Land Conversions," *Econometrica*, January 1961, pp. 1–23; and Paul F. Wendt and William Goldner, "Land Values and the Dynamics of Residential Location," in *Essays in Urban Land Economics* (Los Angeles: University of California, Real Estate Research Program, 1966), pp. 188–213.

fringe site earns no rent at present, but is expected to do so in the future. This expected rent will support a future capital value of the land, which has a capital value now as a function of the rate of discount. It is possible to estimate what the price or capital value must become at some future date, and with a given interest rate to support the current land price appreciation, which has no base in current rental earnings. This is illustrated for a selected number of cities in Appendix Table A–4, using the land value appreciation data generated in Appendix Table A–3. The method can be illustrated as follows: Assume the appreciation over farm and improvement costs is $2,543. At 6 per cent interest this could mean that the expected capital value in five years is $3,403 and in ten years $4,555. This expected value represents the capitalization of future annual rents, which are related to travel savings. For example, if the land is expected to have a capital value of $3,403 in five years, this is the equivalent of an annual rent or travel saving of $266 for twenty-five years capitalized at 6 per cent. In other words, for the property at the end of five years to be worth $3,403 more than its agricultural value and improvement costs, it must save its owner $266 annually in travel costs over the poorest (most distant) homesite then available. Data for several cities are included in Appendix Table A–4.

It would be helpful in formulating land policy to know if the projected future prices of present fringe sites which are necessary to support present

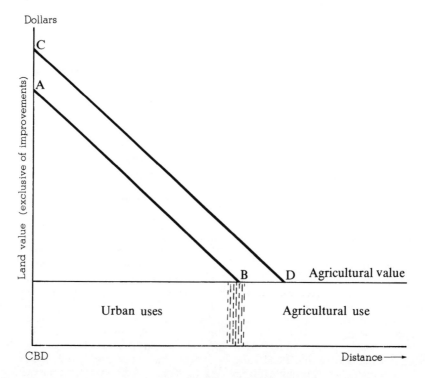

*Figure 3.  Hypothetical land value—distance relationship.*

appreciated prices are related to a reasonable expectation of future travel-saving rents or if other sources must be searched for. Some check on these expected future values can be made by reference to historical trends. It is possible to relate today's prices of fringe sites to today's prices of closer-in sites that have existing houses.

If data were available for ten-year-old developed, close-in lots that indicated their capitalized economic rent value (appreciation above agricultural and improvement costs), this value could be compared with the capitalized rental value of new fringe sites expected ten years hence. If the new fringe sites' expected rent value exceeded the current rent value of existing lots, it would indicate that buyers and sellers expect future developments to move the marginal fringe site in the next ten years more than it actually moved, in terms of travel costs, in the last ten years. This is illustrated pictorially in Figure 4 as follows: Marginal fringe site, *A*, is priced in 1964 so that it represents rent value of $2,116, which is equivalent to an expected capitalized rent value of $3,790 in ten years' time. Now, if the 1964 value of existing ten-year-old lots, *B*, represents a capitalized rent of $2,570, this would indicate that the expected location of the 1974 marginal fringe site, *C*, will be such that it will create a travel saving for site *A* which is greater than the advantage enjoyed by site *B* over site *A* in 1964. In the example, the expected 1974 travel saving of site *A* is $296 annually (for twenty-five years at 6 per cent) or a capital value of $3,790, while the travel saving advantage of site *B* in 1964 is only $201 annually, or a capital value of $2,570.

This type of simplified analysis is made for a number of cities in Appendix Table A–4. It is illustrative only, since no data are available for rent values of existing ten-year-old sites or for the relevant time horizons and discount rate. FHA does have data on site value, *including improvements,* and the average age of these is known to be about ten years.[9] The data on existing lots, then, in Appendix Table A–4 are too high, because they include improvements and are not exclusively capitalized rent values.

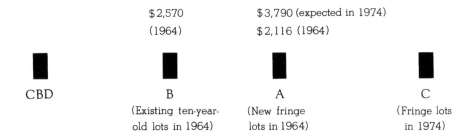

|  | $2,570 | $3,790 (expected in 1974) |  |
|  | (1964) | $2,116 (1964) |  |
| CBD | B | A | C |
|  | (Existing ten-year-old lots in 1964) | (New fringe lots in 1964) | (Fringe lots in 1974) |

*Figure 4. Hypothetical relationship of capitalized rent values of existing and new sites.*

[9] Federal Housing Administration, Division of Research and Statistics, *Statistical Summary,* RR:250, 1964, Table 86.

Also they are of limited comparability, because built-up areas are not exact substitutes for new areas even if they had similar locations, since the type of house is fixed and partially obsolete, or lot size may be smaller and less desirable. However, there may also be some amenity values on the plus side for older established neighborhoods.

While the data are only illustrative, sixteen of the twenty-eight cities that can be analyzed had expected values in ten years which exceeded the current total value of ten-year-old lots, not allowing for development costs. This suggests that developers in some cities are looking forward to more rapid growth in terms of travel savings than has been experienced in the past. If owners actually use a discount rate higher than 6 per cent (which seems likely), then the expected future values are even higher than those indicated in Appendix Table A–4, and even more in excess of the current value of those lots already earning rents.

If this type of data were available, it would be possible to estimate the expected future locations of the fringe building sites (point C in Figure 4) that are built into present fringe land prices. The population growth and densities necessary to realize this location could be estimated. These could be compared to the best available independent estimates of population growth, densities, and transportation costs.[10]

Another type of historical analysis of lot prices can be made. It can be asked whether the rent value contained in the marginal fringe site in year one represents the present rent value actually realized ten years later for sites ten years old. For example, assume a rent value of $813 in 1950. This represents the present value in 1950 of the expected increase in prices of these ten-year-old lots by 1960. This might be computed at 6 per cent interest, and the result would be $1,456. This projected figure could then be compared with the actual rent value contained in ten-year-old lots in 1960. If the actual rent value were higher, it would mean that buyers and sellers in 1950 either underestimated the future prices or used a higher discount rate. However, it is not possible with present data to make such an analysis. For example, 1950 market prices for various cities are available, and so are the prices of sites that averaged ten years of age in 1960, but no improvement cost data are available so that rent value for comparable size and quality might be calculated.

While it is not now possible to test expected rent values, it is possible to ask whether the buyer of a lot in 1950 has found that the total price paid was indeed the then-present value of the lot value he realized ten years later. In other words, it is possible to ask whether 1950 fringe lots represented a good investment when judged by realized prices of ten-year-old lots in 1960. Today's buyers would be expected to be willing to pay existing

---

[10] For an illustration of projected residential and total urban land use, see *Penn-Jersey Transportation Study*, Vol. 2, Philadelphia, September 1964.

prices if they experienced ten years when lot prices rose more than enough to pay 6 per cent interest on the principal.

This type of analysis is made in Appendix Table A–5, and shown graphically in Figure 5. FHA prices of fringe sites in 1950 are used to compute a capital value ten years later at 6 per cent compound interest annually.

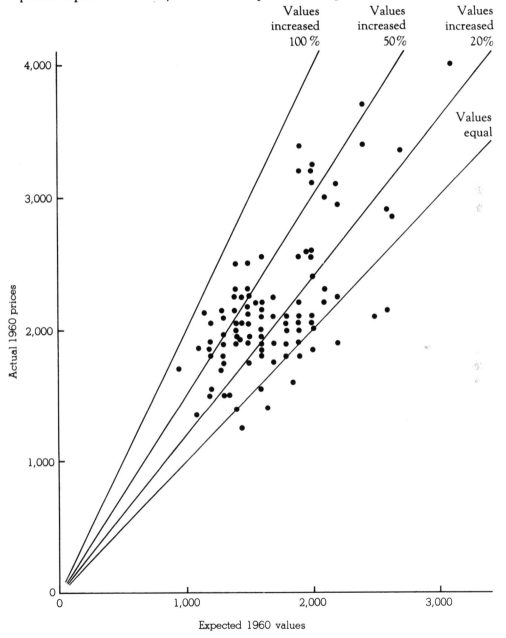

Figure 5. Relationship of actual 1960 prices of ten-year-old lots to value expected in 1960 based on 1950 prices. (Data contained in Appendix Table A–5. Six per cent compound interest was used in calculating the expected values.)

Thus, a capital value in 1960 is computed in such a fashion that the 1950 price is treated as its present value. This is then compared with the price of existing FHA sites in 1960 which are known to have an average age of about ten years. In almost all cities the estimated 1960 capital values have been exceeded by the actual market price of existing lots in 1960. This means that investment in new fringe lots in 1950 earned over 6 per cent judged by realized prices of ten-year-old lots in 1960, though the excess returns have lost some value through inflation.

Another question can be asked about the relative prices of new and old lots. On the basis of rent theory, one would expect existing more centrally-located lots at a given time to have higher values than new fringe sites of comparable quality and size. Appendix Table A–6, using FHA data, indicates the ratio of existing lot prices to new lot prices for various cities. In 1963, out of a hundred cities sampled, only twenty-seven had ratios of existing to new lot prices of more than one. This means that in the majority of cities, the new prices exceeded the prices of older closer-in lots. In comparison, the 1955 ratios are almost all higher than the 1963 ratios and the majority are over one, meaning that existing lots more frequently had prices higher than new lots. This indicates that over the period 1955-1963 new fringe lot prices frequently went up faster than existing lot prices. These data cannot be conclusive, because comparability in lot size and quality cannot be established, nor can changes in transport cost and location of employment and services be enumerated. But it does raise the possibility that fringe site owners expect more future rent increases than are actually realized or expected in established closer-in lots.

Up to now, the discussion has concerned rent, which accrues because of natural spatial conditions that are in limited supply (though this can be influenced by the way population is distributed over the landscape and by transportation technology). However, there are other valuable site characteristics which are to a degree non-reproducible and in limited supply. These include such landscape amenities as unique topography and view, mature trees, and so on. Some areas may be so favorably located as to escape air pollution and smog, or to offer association with a social elite; not everyone can live in the same block with the richest in town.

*Distribution of Rental Gains*

If the currently-observed appreciation in lot values above their cost of production is in fact caused by an expectation of rents that reflects great demands being placed on inelastic land supply, then home buyers face a costly conversion process and the public can only look at alternative ways to distribute the gain—that is, as long as market bargaining rations the scarce supply. Here the ghost of Henry George becomes more active and ideas of capturing for the public the "unearned" rent increment take on new life. These range from outright state ownership of land, as in Eastern

Europe, to state ownership or taxation of the development rights and values, as Great Britain tried briefly between 1947 and 1953,[11] and is implementing again with the Land Commission Act of 1967.[12, 13]

The United States has begun to consider similar measures. President Johnson's proposed Housing Act of 1965 would have provided FHA loan insurance for state agencies to purchase land in advance of development. It was proposed that the loan should equal the acquisition cost of the land and be repayable in fifteen years at 4 per cent interest. Housing and Home Finance Agency Director Weaver said that when resold the land would be priced at "fair value," a term that was not defined.[14] If the land were sold at cost, the new owners would receive an advantage and could look forward to capturing future rents. This would probably necessitate some sort of control over resale. However, if the land were priced at market value, then the state agency would capture the rent for the public treasury.

It is an interesting footnote that the political representatives of agricultural districts have not let this phenomenon go unnoticed. Representative Poage of Texas introduced a 1965 bill which would set up a U.S. Agricultural Land Development Corporation.[15] It is put in the context of surplus crop control, but its effect would be to provide procedures for setting up a joint public-private corporation to buy land in the rural-urban fringe to be sold later at market prices. Presumably, the U.S. Treasury would then get a share of the rent values.

Perhaps we should, before debate on the distribution of rent gains becomes too heated, examine a little further one crucial cause of rent—the central locational advantage. People pay rent to save the expense of living at a distance from value-producing centers of employment, services, and so on. The number of city centers may at first glance seem relatively inelastic

[11] The Town and Country Planning Act of 1947 vested all land development rights in the state. When development permission was granted, a charge was levied calculated on the difference between the present and future use value of the site. It was intended that all land would change hands at existing use value. A fund was set up to meet cases of hardship arising out of the loss of development values. The development charge was abolished in 1953. (For a discussion of the British case, see *Land Values,* being under the Auspices of the Action Society Trust [London: Street and Maxwell Ltd., 1965].) Also see Charles Haar, ed., *Law and Land; Anglo-American Planning Practice* (Cambridge: Harvard University Press, 1964).

[12] *The Land Commission,* presented by the Minister of Land and Natural Resources and the Secretary of State for Scotland to Parliament by Command of Her Majesty, September 1965 (London: Her Majesty's Stationery Office). The Land Commission Act of 1967 will implement a 40 per cent levy on the increase in development value.

[13] For a review of urban land policies around the world see Charles Abrams, *Man's Struggle for Shelter in an Urbanizing World* (Cambridge: M.I.T. Press, 1964), and "Urban Land Problems and Policies," in *Housing and Town and County Planning,* Bulletin 7 (New York: United Nations, 1953). Also see Goran Sidenbladh, "Stockholm: A Planned City," in *Cities (A Scientific American Book)* (New York: Alfred A. Knopf, 1965).

[14] Weaver, *op. cit.*

[15] H.R. 7500, 89th Congress, April 14, 1965.

and thus an inevitable cause of rent. But need it be? Could population growth be directed to new centers, so permitting a substantial saving in total rents?

A relevant phenomenon is now appearing. It is the planned new town. In Great Britain, the new town is well established, but, in spite of tremendous postwar economic growth, the U.S. has till now created few genuine new towns oriented to self-sufficient centers of employment and amenity. But the new community of Reston, Virginia, already exists and Columbia, Maryland, is now beyond the planning stage. They will offer employment, commercial centers, and a wide range of housing types for planned populations of 75,000 and 125,000 respectively.[16] While they are within metropolitan Washington's sphere of influence, the fact that they will absorb population growth has quite different implications for rent values than if these people were added to the Washington fringe and looked primarily to that city's central business district for employment. The new town is being widely discussed now because it offers a higher quality urban environment, but its implications for reduction of land prices should also be explored.[17]

## CONVERSION AND MONOPOLY PROFITS?

The preceding analysis found value appreciations that considerably exceeded opportunity and conversion costs. A method that might determine whether this represented the true present value of generally expected rents was suggested, but data are not now available that would permit firm conclusions. Meanwhile, however, it might be asked whether there is any reason to believe that this appreciation could be something other than the present value of future rents. Is there any evidence or *a priori* reasoning suggesting it could represent monopoly returns as well as rent? Is the land market so structured that gains due to restrictions of supply not dictated by production costs or natural limitations are also obtained?

Observation suggests that land markets are neither dominated by a single seller nor conditioned by fewness or oligopoly.[18] Where there are many sellers organized collusion seems unlikely. Is it then impossible to conceive of returns that exceed conversion costs and the discounted future

[16] Wolf Von Eckardt, "The Case for Building 350 New Towns," *Harper's Magazine,* December 1965, pp. 85–94. Also see Edward P. Eichler and Marshall Kaplan, *The Community Builders* (Berkeley: University of California Press, 1967).

[17] For a discussion of factors affecting aggregate rents, see William Alonso, *Location and Land Use, Toward a General Theory of Land Rent* (Cambridge: Harvard University Press, 1964).

[18] However, one study—of Greensboro, North Carolina, which had a 1960 population of 119,574—showed that two large developers accounted for over 40% of the lots subdivided between 1960 and 1964. See Shirley F. Weiss, John E. Smith, Edward J. Kaiser, Kenneth B. Kenney, *Residential Developer Decisions* (Chapel Hill: Center for Urban and Regional Studies, University of North Carolina, April 1966), pp. 25-26.

value of rents? An analysis follows of those public actions which tend to limit land supply, and of any private restrictions that differ from the usual noncompetitive factors.

*Public Supply Restrictions*

It is conceivable that some government actions contribute to withholding land from the market and therefore to raising prices above cost of production and rent. An example might be the zoning of land for large or small lots. Fiscal considerations such as the need to maximize tax revenues and to reduce the demand for public services control this. Some communities purposely zone available sites only for large lots, hoping thus to reduce government costs in relation to revenues. In the process they put a premium on areas that are open to small lot development. (The effect may be even more extreme in the case of multifamily and apartment zoning which some communities deliberately hold to a minimum or exclude altogther.)[19]

The government may again influence prices wherever it controls sewer and water services, and developers can build only when and where these are available. If the utilities are not developed to keep pace with demand, then supply of sites will be restricted, even if full production costs are charged.

Another source of appreciation above production costs occurs whenever public services are provided for less than what they cost. A portion of the value of a housing site is based on the availability of certain services and characteristics. To the extent that these factors are limited in supply and are made available without or below cost, their value becomes capitalized into the value of the land, to the benefit of the owner at the time they become available.

These features are of mainly two kinds: (1) Characteristics bestowed on the area by expected social relationships; there may, for instance, be a homogeneous group of high income and high status residents. Roger N. Harris, in a study of Raleigh, North Carolina, found this subdivision characteristic significantly related to land values.[20] It might be achieved by such influences as advertising, which do have some production cost to the developer, or by certain kinds of zoning—which have no production cost to the developer. Two other features that Harris found related to high land prices were a predominance of single-family housing in the area and an absence of nonresidential uses in proximate areas. Again, a Los Angeles study testifies to the relevance of such amenity features as neighbors' median incomes, degree of crowding, and percentage of white or nonwhite popula-

[19] See "Apartments in Suburbia: Local Responsibility and Judicial Restraint," symposium published in *Northwestern University Law Review*, Vol. 59, No. 3.

[20] Roger N. Harris, *Determinants of Central Shopping and Residential Land Values*, Ph.D. thesis (North Carolina State University at Raleigh, 1965). Also see Shirley F. Weiss, Thomas G. Donnelly, and Edward J. Kaiser, "Land Value and Land Development Influence Factors," *Land Economics*, May 1966, pp. 230–33.

tion.[21] (2) Another category of site amenity is that made available by public investment, but not paid for by direct charges or taxes on the benefited parties. An example might be near access to a particularly fine school that is not available to other neighborhoods. Or it might be easy access to the limited number of public parks.[22] Only a few areas will offer such values to lot owners, but their tax rates may well be identical to those of less fortunate areas.

The limited provision of certain utilities at less than cost is another example. There is evidence that some cities for various reasons provide utility services to new areas at the same price as to the central city area, even though their cost may be higher.[23] A case occurred recently in Montgomery County, Maryland, where an expensive pumping station had to be built to serve a new area, but the residents there were not to be asked to pay any more for sewer service than was the rest of the sanitary district.

It has been shown that some features of a house site are made available without or below cost. If these are in fact in limited supply, though provided at constant cost, they will become capitalized into land values as people bid to get access to them. A house site buyer will not have that access until he owns land where they are available, so the land buyer can charge for amenities which cost him nothing to produce. The local government might have hoped that if services were provided without or below cost the price of lots would be correspondingly low.[24] But the saving may not be passed on to the final lot consumer, and the improvement merely increases the property owner's gains.

A question for research is how much of the asset appreciation which accompanies the land conversion process is based on rent reflecting the value of limited amenities provided at less than cost. If the public wishes to provide certain amenity values below cost it could avoid their capitalization into land values by providing them in such a manner as not to restrict their supply more than is dictated by rising cost functions. In other words, it could avoid adding to the scarcity of a given quality of land by not restricting the supply of any amenity feature which is reproducible.

This asset appreciation reflecting the value of amenities provided in

[21] E. F. Brigham, *A Model of Residential Land Values* (Santa Monica, California: The RAND Corporation, Memorandum RM–4043–RC, August 1964). The same material appeared in article form in *Land Economics*, November 1965, pp. 325-34.

[22] See Jack L. Knetsch, "Land Values and Parks in Urban Fringe Areas," *Journal of Farm Economics*, December 1962, pp. 1718–26.

[23] For a discussion of this, see Mason Gaffney, "Containment Policies for Urban Sprawl" in *Approaches to the Study of Urbanization* (Lawrence: University of Kansas Press, 1964).

[24] Public policy on the extension of paving, water, and sewer services to new developments has other objectives, of course, such as the location of new building. For a case study of this kind of impact, see Shirley F. Weiss, John E. Smith, Edward J. Kaiser, Kenneth B. Kenney, *Residential Developer Decisions, op. cit.*, p. 27.

limited supply at less than cost appears as a rent from the developer's point of view, but is monopoly from the point of view of the whole economy, in that it results from a contrived rather than a natural restriction of supply.

*Private Supply Restrictions*

It was mentioned above that gains based on supply restrictions by private individuals and firms do not appear to be common. While information on market structure is limited and more would certainly be useful, from what we have it does not appear that monopoly or oligopolistic practices are widespread. Is there then anything else in the operation of private parties in the land market that creates the effect of a monopoly-like supply restriction?

It will be argued here that there is no *a priori* reason that the reservation or asking price set by sellers should not be found to exceed the present value of actual future values, and no reason that this price could not persist over considerable time, even if there is no overt collusion and no domination by a few large sellers. As sellers seek to establish the market price, they set certain prices in a given year and see what rates of sale they obtain. This reservation price is the present value of the price expected at some date in the future. Suppose, as shown in the left-hand chart of Figure 6, that at some future date (ten years)[25] most sellers expect the price to be $P_x$. Today's price then is the present value of $P_x$ using an appropriate discount rate (and including other net holding costs such as taxes).[26] Suppose this turns out to be $P_1$. Given the demand curve (for year one) for the land quality being considered, this experimental price will produce a certain rate of sale. Acreage $OA$ out of a total supply of $OC$ will be sold in year one. Owners will be indifferent between selling $OA$ in the year one at $P_1$, or keeping all or a portion of the available acreage and selling it for $P_x$ at whatever future date was used above. The reservation price in year two will have to be slightly higher than $P_1$ to maintain this indifference between future and present prices.

In this example, if demand in year two and year three remains approximately what it was in year one, then at the end of year three all of the available acreage of the given quality will be sold, at $P_1$ and slightly above. Is there no day of reckoning eventually? At the end of ten years it will surely be known whether the expected price is about to be realized, and it may turn out that sellers were unduly optimistic. However, by that time all the land of that quality will long have been sold at the seller's reservation price. The sellers, though now proven wrong, have all realized their profits, and it is the lot buyers who find they have paid more above production

[25]The future date used is somewhat arbitrary. It depends on the planning horizon of the sellers, but in any case is limited by the fact that, with discounting, very distant gains are largely irrelevant.

[26] Holding costs, taxes, and risk are discussed in detail in a later section.

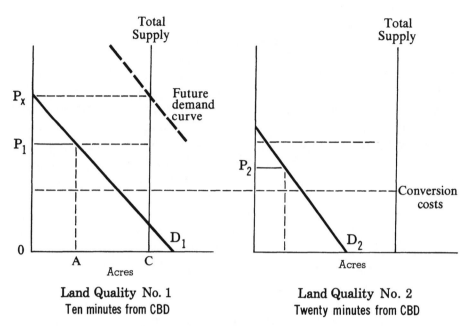

*Figure 6. Hypothetical relationship between present and future values.*

costs than the rental value of the land. With the usual storable product, the case is different. For example, owners of a corn crop in year one may set a reservation price which reflects their judgment that there will be a short crop next year. If they are right, they will make a profit, but there will be a new crop one year later, and if they have made a wrong estimate they will lose, because the average price for the two years is less than their reservation price. In the case of land of a given quality it is possible for all the sellers to realize their reservation prices before the actual price of the existing stock in the future is known. Only a shift in the demand curve for the yet unsold land can hurt an owner. If the reservation price is too high, the rate of sale will be slow and some land will be left unsold as the projected date approaches. If yearly demand does not increase or at least remain constant, many owners will see that the actual price in year ten will not be as expected; some will try to sell quickly by lowering their asking prices, and the market will break. Success in realizing the reservation price set at the outset depends upon a rate of sale that will clear the market before it can become evident that the future price was estimated too high.

However, competition between sellers to be the first to sell is not strong enough to affect price. It might be asked why random experimentation with lower prices by a few owners or differences in expectation and discount rates would not break the market at the outset, as is usual with a reproducible product in a competitive market. If a few sellers were to undercut the prevailing reservation price, they would be sure to sell their land at an early date. But because they do not represent a productive capacity which would create a supply at the same price in the future, they can be ignored by the rest of the owners. Only if many were to sell and shake the confidence of most sellers would the price be forced down. So the controls on price usual in competitive markets are not necessarily present in the land market. If the rate of sale is reasonably rapid, all sellers of a given quality of land realize their reservation prices, and there is no competitive pressure created by fear of being the last seller, because the projected date is still far enough away that the expected price remains unquestioned and unproven. And there is no pressure from a few sellers who might experiment with lower prices. In products with some elasticity of supply such experimentation reduces the market price, but such is not likely to be the case with land.

Now, it may be asked if buyers must share this over-optimism if the above conditions are to hold. The answer is that they need not necessarily. A buyer may expect a future price which would be too low to support the present value asked by sellers. But he is like anyone who buys from a monopolist. If he values the lot highly enough as a consumer good to justify its present cost, then he will buy it even if he judges that it is not an attractive investment good whose present value will be supported by future rents. Of course, lots will still sell more slowly than they would if the seller's expected future prices and their present values were lower; when buyers expect reservation prices to decline, most of them will be motivated to wait and buy later. But many have no recent experience of declining prices, and, though it is unknown how many lot and home buyers make any estimate of what future travel savings or other rent-creating factors are implicit in present asking prices, if they do not expect decline they will buy—even if they judge that future rents will not support the price. They need only sufficient demand, that is, the income and the necessary tastes.

The above analysis assumed that the buyer had reasonably good price information. But there is also a typical final purchaser who buys land and buildings in a package, where the land price may not be clearly evident.

The situation suggests there may be a need for a land market news service that will stiffen buyer resistance to prices which appear to lack the support of future rents. Current sales and locations might be listed, along with growth rates and transportation costs, and projections might be made. The individual buyer could then take this expert opinion into account when buying, and sellers' gross optimism should be thereby reduced.

One final possible means of correcting excessively high reservation prices has to be considered. This is activity on the part of sellers of other land. For example, in the cases illustrated in Figure 6, we looked first at land ten minutes from the CBD. Does the existence of land *twenty* minutes from the CBD constitute a corrective to the reservation prices of the first land? The answer is that it does not if its relative reservation price reflects the travel differential. The demand curves for each land quality or distance zone depend on the prices for all other land. But there is a possibility that the whole structure of prices is too optimistic, while the relative prices correctly reflect the distance and travel cost differentials. So it seems unlikely that the buyer can look to a more distant land area for a better buy relative to its distance disadvantage.

It seems logical to conclude that in a dynamic, growing land market it is not production costs that determine the present supply price but people's expectations of future prices and rents.

Owners could build up expectations of future prices based on projections taken from historical experience, without giving any thought as to whether in this case the increases could reasonably be supported by increased rents reflecting travel savings or other items of inelastic supply. These expectations could then support a widely-held present value reservation price which was not necessarily a conscious withholding of supply to achieve a certain price above present values of future rents, but which had the same effect.

If these expectations finally turned out to be wrong, profits garnered along the way might still not be affected; while some of the hold-outs at the end might lose money, the early profit takers would be safely home. So even if expectations are wrong, they will effectively bring returns above costs and realized rents, so long as they are commonly shared. Even if the past has shown that rent expectations may not be met, future outlooks need not be dampened.

Rent value can only be determined *ex post facto,* for one can never be sure that the return now called "present value of expected rents" will not turn out to be in fact a profit. The important thing to note is that a mistaken estimate will mean merely that the return the owner once regarded as rent will turn out to be profit. A commonly shared overestimation can mean high profits.

I would hypothesize that two factors are now operating to give owners mistaken views of future prices and rents:

1. Some of the value of mature areas is amenity, and not the travel-saving location. This amenity value will not automatically accrue to today's new area in the future just because the rural-urban margin extends itself.

2. As a city grows, there is more acreage in each one-mile-deep ring succeeding out from the core. If one forgets this, one gains a false impression from past experience with population growth and household formation of what a given absolute gain in population will mean in terms of the location of the fringe site.

No data are available here to prove that sellers are in fact overly optimistic at present or were in the past. Yet, it has been shown that this is a relevant research question since there is no *a priori* reason to expect that a bad guess about the future will not continue for a number of years, with resulting higher monopoly-like prices to many consumers.

## FACTORS AFFECTING LANDOWNERS' ABILITY TO WITHHOLD SUPPLY

An important theme of the above analysis of the land market centers around the ability of landowners to wait and hold land to realize rents and profits. Reservation prices depend on the rate at which future expected prices are discounted. The higher the discount rate the lower present values will be. At some point the discount rate can be large enough to drive present values to zero, and then the supply price will be simply the conversion cost of the land, which consists of development costs plus farm values.

The discount rate depends upon the opportunity cost, or interest rate, of capital plus other costs such as taxes[27] and the uncertainty factor. Both of these will be referred to here as the total yearly holding cost rate.

### Taxation

Under present tax rules property taxes are deductible from federal income tax. This means that a 2 per cent property tax for someone whose marginal income tax rate is 50 per cent pays only the equivalent of a 1 per cent property tax. Other rates are shown in Table 14. The effect of this is that for many land investors increase in the property tax does not raise their holding costs accordingly.

Table 14.  Effective Property Tax Rates for Various Marginal Income Tax Rates

| Marginal income tax rate—per cent | Holding cost for a property tax of 2 per cent | 4 per cent |
|---|---|---|
| 66 | 0.66 | 1.3 |
| 50 | 1.00 | 2.0 |
| 33 | 1.33 | 2.6 |

To illustrate the effect of increased property taxes, assume first that the interest rate is 5 per cent and property taxes are 2 per cent compounded. If the owner is in the 50 per cent marginal income tax bracket, this means a 1 per cent property tax rate for a total holding cost of rate of 6 per cent. Further, assume that farm value amounts to $1,000 per acre and it is expected that the future value in ten years will be $2,159, an appreciation of about 116 per cent in total, which is a compound rate of 8 per cent annually. At a 6 per cent holding cost rate, the present value of $2,159 in ten years is approximately $1,205, which is above farm value and becomes the reservation price. Or to put it in other terms, the $1,000 investment in land, which has an expected gross value of $2,159 if held for ten years, is

[27] For an annotated bibliography of tax effects, see John Rickert and Jerome Pickard, *Open Space Land, Planning and Taxation: A Bibliography* (Washington: Urban Land Institute, February 1965).

better than earning 5 per cent net compound interest for that period.[28] These values are shown in Table 15. If the reservation price exceeds the market price, the owner will hold the land for future gains.

However, when property taxes are raised to 6 per cent[29] and total holding costs become 8 per cent, the present value drops to $1,000; the owner breaks even on his investment and the reservation price equals the conversion costs. The owner is indifferent between investing $1,000 in holding this land or making another investment yielding 5 per cent net compound interest. In ten years the gross value of $2,159 has a net value of only $1,629 after taxes, which is the value of $1,000 in ten years at 5 per cent compound interest.

At still higher property tax rates, it becomes unprofitable to invest in land to hold for appreciation. It should be noted, however, that in this example the property tax had to be raised from 2 to 6 per cent (and the total holding cost from 6 to 8 per cent) to make land holding unprofitable when total appreciation of 116 per cent was expected in ten years. This amount of appreciation has been exceeded in many places in the U.S. To generalize from the above example, Table 16 has been prepared to show the required ten-year total appreciation necessary at various total holding cost rates to break even. For example, when appreciation is about five times or 500 per cent it takes a holding cost rate of 20 per cent to reduce present values to zero and prevent land holding for future gain.

It should also be noted that while a property tax increase can lower prices to lot consumers it may not necessarily reduce the amount of appreciation above farm value, since if the property tax rate increase is general, the price of agricultural land could also be expected to fall.

The landowner's financial situation is important in assessing tax effects.

Table 15.  Present Value of Land Worth $2,159 in Ten Years at Various Compound Holding Cost Rates and the Value of a $1,000 Investment at Various Compound Interest Rates

| Holding cost and discount rate | Land price | Value of $1,000 at end of year: | | | | | | | | | | Present value of $2,159 |
|---|---|---|---|---|---|---|---|---|---|---|---|---|
| | | 1 | 2 | 3 | 4 | 5 | 6 | 7 | 8 | 9 | 10 | |
| (per cent) | (– – – – – – – – – – – – – – – – – – dollars – – – – – – – – – – – – – – – – – –) | | | | | | | | | | | |
| 5 | 1,000 | 1,050 | 1,103 | 1,158 | 1,216 | 1,276 | 1,340 | 1,407 | 1,477 | 1,551 | 1,629 | 1,326 |
| 6 | 1,000 | 1,060 | 1,124 | 1,191 | 1,262 | 1,338 | 1,419 | 1,504 | 1,594 | 1,689 | 1,791 | 1,205 |
| 7 | 1,000 | 1,070 | 1,145 | 1,225 | 1,311 | 1,403 | 1,501 | 1,605 | 1,718 | 1,838 | 1,967 | 1,097 |
| 8 | 1,000 | 1,080 | 1,166 | 1,260 | 1,360 | 1,469 | 1,587 | 1,714 | 1,851 | 2,000 | 2,159 | 1,000 |

[28] The increase in taxes causes present values to fall. This lower asking price may cause an increase in the rate of sale of the total supply of land of a given quality, e.g., the demand may have been such that all available will be sold in six years at the higher asking price but in three years at a lower asking price. The rate of sale cannot be predicted unless the demand curve is known.

[29] Actually the equivalent of 6 per cent compounded.

Table 16.   Per Cent Total Appreciation Needed to Break Even in Ten Years
at Various Yearly Holding Cost Rates Compounded

*(Per cent)*

| Holding costs | Total appreciation |
|:---:|:---:|
| 6 | 79 |
| 7 | 97 |
| 8 | 116 |
| 9 | 137 |
| 10 | 159 |
| 15 | 304 |
| 20 | 519 |
| 25 | 831 |
| 30 | 1,279 |
| 40 | 2,793 |

Not only is his income tax bracket important, as indicated above, but if he is in a situation of cash scarcity and cannot borrow easily, the impact of high taxes is greater.[30]

To conclude, the impact of a property tax increase (or increase in other holding costs) is to reduce the reservation prices of land owners, but the increase will have to be sizeable to prevent land holding for future gains if expected appreciation is large. It would appear that there are definite limits to the use of property taxes to substantially reduce present market prices for land, unless fringe land is taxed at discriminatory rates. Also the property tax rate must be continued even after the lot is sold to consumers, since a windfall would result if taxes declined and the land value continued to increase.

It should be noted in passing that the effect of a lowering of property taxes (other things being equal) will be to raise present values and reservation prices of fringe land. Preferential tax assessment systems such as are found in Maryland[31] could be expected to increase the asking prices of fringe land and to increase landowner gains.

The above discussion is cast in the framework of property tax effects on the holding of land in anticipation of price increases, and the tax effects on prices to lot buyers. But at several points in this study reference has been made to taxation which could shift part of the appreciation to the public treasury. The property tax could not only have an impact on current land prices, but also serve to capture part of the gain for the public. Another possibility should be noted—a tax on capital gains, in this case the

[30] For an empirical example, see Kenneth B. Kenney, *Pre-development Land Ownership Factors and Their Influence on Residential Development* (Chapel Hill: Center for Urban and Regional Studies, University of North Carolina, Sept. 21, 1965), mimeographed, p. 27.

[31] See Peter House, *Preferential Assessment of Farmland in the Rural-Urban Fringe of Maryland* (Washington: U.S. Department of Agriculture, ERS–8, June 1961). Also see *Farm Land Assessment Practices in the United States* (Chicago: Council of State Governments, 1966).

net appreciation. Under current federal income tax procedures, capital gains are more favorably treated than regular income. When applied to land, the procedures create varied results but, in general, people engaged in land improvement and development pay regular income tax on land sales, while speculators get the more favorable capital gains treatment, unless land sales are their major source of income.

A special flat rate capital gains tax on land might be designed that would take a given portion of the realized appreciation.[32] Since the capital gains tax occurs after the appreciation has occurred, it does not affect holding costs. So imposition of a capital gains tax or an increase in its rate would not reduce lot prices to home buyers,[33] but it could capture part of the appreciation for the public treasury. As noted above, property taxes increase holding costs and reduce present values. However, capital gains taxes are not holding costs but, rather, occur when gains are made, whether at a current or at a future sale.

As the flat rate tax on capital gains was increased the attractiveness of land holding relative to other investments would decline and at some point speculation would be reduced. Also more land would be held by active farmers, who receive a current agricultural income flow until such time as the land is actually in demand for urban building.

## Capital Costs and Risk

The discussion of the factors affecting landowners' ability to withhold supply can now be generalized to include the other major component of holding costs, namely the cost of capital, which is reflected in the discount rate. Casual observation of the land market indicates that it is a relatively high risk business with relatively high rates of interest on borrowed capital. In fact, some conservative financial organizations will not lend at all for land purchase.

While this analysis has suggested significant appreciation accompanying the conversion process, this does not mean that all land speculators and developers receive large rents or profits or even that the average rate of gain realized by developers is high. While gains may be large in total, in the individual case there is great uncertainty as to the timing of the price increase, and the net gain to the developer or speculator may be eaten up in interest charges if predictions prove erroneous.

The development of a particular parcel depends on a host of factors such as consumer tastes, the extension of public facilities, zoning decisions, rate of population growth, and so on. These vary in predictability and

---

[32] For a brief discussion of land value increment taxation, see Dick Netzer, *Economics of the Property Tax* (Washington: The Brookings Institution, 1966), pp. 212–13.

[33] If speculative activity were to be reduced it would reduce the market prices of land not currently demanded for immediate use, but it would not necessarily reduce selling prices for active development for urban use.

impart high risk to capital invested in conversion. Some elements of risk, such as change in consumer tastes, are an accepted part of market trans- actions and some exposure to them is desirable and unavoidable. Other elements of risk, however, are features of a particular market organization, and research might be directed to discovering means for their elimination. Better market information and planning of public decisions might reduce the risk and lower the discounting of future values. This would in general have the effect of increasing present values of any given expected future value. Current prices would become more like future prices. A reduction in risk and consequently in the discount rate is similar in effect to a reduc- tion in property taxes—namely, it increases present values. Thus, it can be expected to affect the distribution of the appreciation gain and not its eventual size. For example, the original farm owner might be expected to realize more of the appreciation if its occurrence were more predictable. The farmer is less likely to get caught by problems of capital rationing and cash shortage with lower rates of interest, and will thus be less susceptible to offers by speculators with superior financial positions. With a decrease in taxes or the discount rate the present value increases and the landowner captures more of the eventual appreciation, which before had been taken by interest or tax payments or captured later by the house and lot buyer.

## TAXATION OF APPRECIATION VALUE
## AND MARKET FUNCTIONS

In principle, there seems to be no reason why holding costs could not be increased to the point where the supply price of land would become the agricultural value plus development costs, with no reflection of future appreciation in present prices. Holding costs could be raised by property taxes. Also capital gains taxes could be designed to absorb the total apprecia- tion. A question must be raised at this point on how this would affect the functions now performed by the land market. One function of a rise in price that creates return above costs is to indicate that consumers want more acres of land for lots of a given quality (say ten minutes from the CBD). Since the total supply is inelastic the signal is of no importance for this function. Another closely-related function is to indicate a demand for a change in land use from rural to urban. If all gains above agricultural values were taxed, then developers could only buy land as it came on the agricul- tural market through farmer death and such like. If a faster rate of transfer were demanded a return sufficient to encourage active farmers to move would have to be allowed, and established by market bargaining.

Another function of the rise in price in the market is to ration the available supply of a given quality of land among the possible buyers. If all of the appreciation is taxed away, the land owner or developer will sell all he can whenever the farm opportunity and development costs can be recovered. However, because the consumer can see that the land under consideration may have a locational and transportational advantage in the future, there will be more buyers at the nonappreciation price than there is land of that quality. Unless the capital gains of consumers are also taxed, there will be pressure on the part of the fortunate buyer to resell. Even if the gains were taxed away from all parties, if black market and side payments could be policed, and if sale were on the basis of first come, first served, a problem would remain. This is the question of allocation of the different qualities of land to those who "value" it the most. Unless we want to resort to some kind of governmental rationing system, rents must be allowed to develop and some percentage of these must accrue to the landowner if the rationing function is to be performed. This sort of idea may lie behind the Land Commission Act of 1967 in Great Britain which implements a 40 per cent levy on the increase in development value in contrast to the earlier experiment with a 100 per cent charge in the Town and Country Planning Act of 1947.

If public policy were to attempt to reduce land prices to the consumer, the problems of transition would be gigantic. I have suggested various alternatives here, such as taxation of gains, control of zoning power, reduction of the influence of fiscal considerations in land use planning, improved transportation, better timing and distribution of public services, better land market information, and encouragement of new towns. If these were successful, they would tend to bring down the cost of new sites and existing stock as well. Such a devaluation of assets has wide side effects. As Edward Eichler has suggested: "It may now be that the economy, particularly in California, is so precariously dependent upon an assumption of real estate inflation that the readjustments would turn into an uncontrollable downward spiral."[34] Such is the problem that confronts social science research.

## FACTORS RELATED TO APPRECIATION LEVELS: AN EMPIRICAL TEST

Now that we have discussed some of the factors which might affect the level of appreciation, we can return to the variability in appreciation among cities that was observed earlier, and make some tentative empirical tests to explain it. This will be done first by a simple correlation analysis.

As already noted, data on some of the possible contributing factors are not available and for others the variables can only be proxies. With reference to factors which might be related to the capitalization of expected rents, data are available on the rate of change in population for the city and for the whole urbanized area and on change in land area for the urbanized area between 1950 and 1960. Also, data are available on the distribution of the population. These include the population density of the urbanized area,

[34] Edward P. Eichler, comments contained in Bernard Weissbourd, *Segregation, Subsidies, and Megalopolis* (Santa Barbara: Fund for the Republic, Center for the Study of Democratic Institutions, Occasional Paper No. 1 On The City, 1964), p. 18.

the percentage of the entire urbanized area's population that lives in the fringe, i.e., outside the political boundaries of the central city, and the percentage of the fringe population living in the SMSA ring outside the political boundaries of the central city that works in the central city.

Median family income data are also available, which might indicate differences in lot size and quality, or might indicate that monopoly-like elements extract gains according to the incomes of the area. Finally, several measures of absolute size have been tested. These include total population of the urbanized area and of the central city as a political subdivision, and also total land in the urbanized area.

The analysis was made with the percentage net appreciation above farm value contained in raw land prices as the dependent variable. (Data are contained in column 4 of Appendix Table A–7.) A similar analysis might have been made on the finished lot price data, but it was judged that at present the price and cost data are not sufficient for such analysis. For the interested reader, simple correlation results are also shown in the following table for the absolute dollar amounts of net appreciation, but text discussion will be limited to the percentage net appreciation.

Relationships were tested for 1960, since that is the only recent year for which census data are available. It should be noted that data are used mostly for the urbanized area, as defined by the census, rather than for the SMSA. It is judged that the urbanized area is the most consistent area for comparative analysis, since the SMSA is affected greatly by the accident of county size, which varies markedly across the United States, and particularly affects average income and population per square mile.

The urbanized area includes at least one city of 50,000, as well as the surrounding closely settled areas which, in the case of unincorporated areas for example, must have a population density of at least 1,000 per square mile.

Some data on the primary city as defined by the political boundaries are included also,[35] but conclusion from these must take into account variability of annexation policy and aggressiveness, which may mask real differences in central city characteristics defined by functional criteria rather than by political lines.[36]

The raw data for the independent variables are shown in Appendix Table A–9, and the simple correlation coefficients and levels of significance

[35] Variables for both urbanized area and central city were included because the fringe growth sometimes takes place within the political boundaries of the central city as well as outside them. Also, it is of interest to test relationships between central growth and total urbanized growth. This, however, creates statistical problems of multicollinearity. Variables for the urbanized area and central city are correlated with each other. This does create uncertainties about the statistical results, so they must be treated with caution.

[36] This is not to suggest that annexation policy may not be a factor in price appreciation, but only that these data do not measure it.

are shown in Table 17. The highest simple correlations exist between the population change variables and the percentage appreciation in raw land prices in 1960. The percentage change in population of both the urbanized area and the central city between 1950 and 1960 had simple correlation coefficients of .38 which were statistically significant at the .005 level. The growth rate of the housing areas seems to be positively related to the level of appreciation. This means that areas that are growing rapidly may have difficulty keeping appreciation down. However, the exceptions are worthy of note. A number of cities had higher than average change in population and still managed to have appreciation levels below average. These cities include, for example, Fresno, Sacramento, Wilmington, Jacksonville, and Wichita; others may be noted from Appendix Tables A–7 and A–9 and deserve more intensive study to discover which of their characteristics and public policies contributed to low appreciation in spite of high growth rates.

The other growth rate variable of percentage change in land area, 1950-1960, was also significant and had the next highest coefficient of simple correlation.

Both the median family income variables for the entire urbanized area and the city were significant.

Only one of the population distribution variables proved significant. This was the percentage of the urbanized area population that lived in the

Table 17.   Simple Correlations between Suburban Raw Land Price Appreciation and Selected Area Characteristics, 1960

|  | | Simple correlation coefficients | |
| --- | --- | --- | --- |
| Variable | Means | Per cent appre-ciation | Absolute appre-ciation |
| 1.  Per cent appreciation over farm value | 1,466 | 1.00 | n.a. |
| 2.  Absolute appreciation over farm value | $1,811 | n.a. | 1.00 |
| *Data for urbanized area* | | | |
| 3.  Per cent change in population, 1950–60 | 35 | .382[a] | .394[a] |
| 4.  Per cent change in land area, 1950–60 | 94 | .254[a] | .108 |
| 5.  Median family income | $5,935 | .156[a] | .298[a] |
| 6.  Per cent population living in fringe | 31 | −.167[a] | .169 |
| 7.  Land area (sq. mi.) | 128 | .053 | .540[a] |
| 8.  Total population | 454,176 | .020 | .543[a] |
| 9.  Population per square mile | 3,349 | −.050 | .143[a] |
| *Data for city (political boundary)* | | | |
| 10.  Median family income | $5,667 | .232[a] | .326[a] |
| 11.  Per cent change in population, 1950–60 | 29 | .388[a] | .278[a] |
| 12.  Total population | 180,498 | .070 | .330[a] |
| *Data for SMSA* | | | |
| 13.  Per cent population living in SMSA ring that work in city | 38 | .005 | −.123 |

n.a. = Not applicable.
[a] Significant at the .05 level or less.
*Sources:* See sources to Appendix Table A–9.

fringe, i.e., outside the political boundaries of the main city. This had a negative relationship, with the appreciation level falling as the percentage of the population living in the fringe increased.

The other population distribution variables, of percentage of the population living in the SMSA ring who worked in the main city and the population per square mile for the urbanized area, were not significantly related to appreciation. The measures of absolute size of the population and land area also proved insignificant.

The above simple correlation analysis was done on samples of various sizes according to the extent of the data. In order to analyze some of the interrelationships of the variables and their relative importance, the following multiple regression analysis was made. The sample size of 130 differs from the above analysis in that the smaller cities—generally under 50,000 population in 1950—were dropped because of lack of data. The percentage net appreciation of raw land prices above farm values averaged 1,283 per cent for this group of 130 cities, while it was 1,466 per cent for the group of 260 which included smaller cities. (Mean city 1960 population for this sample of 130 cities was 303,872, while for the whole group of 260 cities it was 180,498.)

The regression results are shown in Table 18. The coefficient of multiple correlation ($R^2$) was .41, which was significant. The analysis indicates that the variables for the urbanized area of land area (square miles), total popula-

Table 18.   Multiple Regression Analysis of Suburban Raw Land Price Appreciation, 1960

| Variable | Regression coefficients | Beta weights | Significance level | Partial correlation coefficients |
|---|---|---|---|---|
| 0.  Constant | 746 | | | |
| *Data for urbanized area* | | | | |
| 3.  % change in pop. 1950–60 | − .493 | − .013 | .87 | − .011 |
| 4.  % change in land area 1950–60 | − .932 | − .065 | .51 | − .061 |
| 5.  Median family income | − .129 | − .080 | .71 | − .035 |
| 6.  % pop. living in fringe | 3.369 | .061 | .60 | .049 |
| 7.  Land area | 5.044[a] | .886 | .00 | .302 |
| 8.  Total population | − .001[a] | −1.272 | .00 | − .272 |
| 9.  Pop./sq. mi. | .234[a] | .214 | .06 | .177 |
| *Data for city* | | | | |
| 10.  Median family income | .013 | .008 | .92 | .004 |
| 11.  % change in pop. 1950–60 | 16.857[a] | .550 | .00 | .379 |
| 12.  Total population | .001[a] | .610 | .04 | .188 |
| *Data for SMSA* | | | | |
| 13.  % pop. living in SMSA ring that work in city | −7.029 | − .077 | .30 | − .094 |

*Notes:* Observations, 130. Numbering of items corresponds with that in Table 17.
   [a] Significant at the 0.06 level or less. Means and standard deviations of the variables are shown in Appendix Table A–9.

tion, and population per square mile, were significant, along with the percentage change in city population 1950–1960 and total city population within political boundaries. All of the significant variables were positive except total population for the urbanized area, which had a negative sign. The analysis of the relative importance of the significant variables depends on the question asked. In terms of reducing the $R^2$, the deletion of the percentage change in city population 1950–1960 would have the greatest effect.

The relationship between the change in city population 1950–1960 and appreciation is consistent with commonly-made hypotheses. The lack of significance for change in population for the whole urbanized area, however, is puzzling.[37] Reasons for the importance of the absolute size variables of urbanized land area in square miles, and total population of the urbanized area and the city, are not readily apparent. Nor are the reasons apparent for the fact that of these total size variables which might be expected to exert their influence in the same direction, the urbanized area total population variable has a negative sign while land area has a positive sign as does city total population.

The above analysis can only be considered as tentative and exploratory. Much more work is needed on cross-sectional and time series data, with both linear and curvilinear models. Data must be developed to get behind the differences in urban physical characteristics to differences in public policy and the private practice and organization of land developers. Further research on a case basis should also be fruitful, for the understanding of the exceptions to the rule may teach us as much as overall relationships.

---

[37] The simple correlation between percentage change in urban area population and in the city population was 0.68. See fn. 35.

# IV

# CONCLUSIONS AND
# POLICY IMPLICATIONS

This monograph is directed toward the problems of developing concepts relevant to the conversion of land from rural to urban use and at analysis of the conversion processes these concepts suggest. Some of the policy implications of the data analyzed above have already been noted and will be extended here, along with suggestions for future research. In conclusion, it seems wise to admit that the present state of scientific knowledge does not permit the formulation of a definitive outline of the complete set of processes involved in land conversion. The principal objective of this study, therefore, is to develop those parts of the outline that are evident from the empirical analyses of the present study, to stimulate thought, provoke discussion, and encourage further work.

The main emphasis of the monograph is to call attention to the appreciation in lot prices above costs of production and farm opportunity costs as a subject for inquiry. Most of the research to date on urban land values has focused on explaining differences in total prices between cities or price gradients within a given city. The evidence presented here supports the conclusion that appreciation levels are significant and measured in the hundreds per cent. Two data sources provide estimates of an average appreciation in land values above farm prices (less improvement costs) of 892 per cent and 1,875 per cent for selected U.S. cities in 1964. Even if future research should revise these estimates downward, there is nothing to suggest any probability of this appreciation being reduced to insignificance.

A simple correlation analysis showed that the greater the percentage change in population growth (of the central city and urbanized area) the greater the percentage appreciation in land value. However, the fact that there is considerable variation in appreciation levels among cities, even among those whose populations are all growing rapidly, suggests a fruitful area for further research. The characteristics and public policies of those cities which are growing rapidly and still have less than average appreciation may indicate possibilities for other areas. Some of the other variables significantly associated with price appreciation were percentage change in land area, median family income, and percentage of population living in the fringe.

There is a paucity of basic data on prices and such items as improvement costs and lot size. The contrast between urban and agricultural land price and production cost data is striking. Investment in national statistics making possible comparisons among cities should have a high payoff. The resolution of many of the questions raised here awaits the development of additional data.

Research should be directed to the components of the finished lot prices. Studies of production efficiency and economies of scale in development of finished lots, as well as of firm organization and financing problems, are needed.

The problem of economic rent paid for transportation advantage or other items of inelastic supply has long occupied scholars. But additional work is needed on methods which might reduce rents, such as new towns and other transportation cost-saving technologies and arrangements.

Appreciation in land values above costs of development and farm values seems to be large and growing larger, on the basis of the approximate and preliminary analysis made here. Further quantification and elaboration are needed. An unresolved question centers on what population growth and extension of the rural-urban fringe is assumed to support appreciation values which constitute the present value of expected future rents. Estimates should be made of the future population growth, densities, and transport costs necessary to support present values. These should then be compared to independent estimates of these rent generators. The current appreciation in land values may simply reflect anticipated economic rent values to be realized in the future. However, the analysis contained herein begins to raise questions as to whether returns exceed production costs and a reasonable expectation of transport cost-saving rents. Past relationships are a limited guide, but there is some evidence that there is considerably more expectation of future value increases built into the prices of current fringe sites than that actually realized in the history of established closer-in lots or than the current market of these older lots recognizes. While the issue remains unresolved, the existence of profit as well as rent seems worthy of further exploration. One possible source of profit returns is the process of speculation. Research is needed on the role of price leaders and how expectations are formed and communicated. Other private market factors which affect the rate of sale and development of building sites need study, along with public policies like taxation.

While attention has been traditionally focused on transportation cost differentials, other factors may produce rents, or profits. These include various amenity factors and differentials in public services and utilities. It needs to be established how much appreciation is based on the value of amenities and other land features which, while reproducible, are nevertheless limited and differentially supplied, for various reasons, in practice. Certain practices of local governments may limit land availability and the rate of development, such as zoning procedures, forms of local taxation, and the provision of utilities.

## IMPLICATIONS OF ECONOMIC RENT AND ITS DISTRIBUTION

Economic rent and the so-called unearned increment have long drawn the attention of economists and would-be reformers. While rent seems to provide a sizeable addition to the cost of a home site and is a source of income transfer, its real significance may lie in its effect on the implementation of public land-use policies. Many suburban environmental features are collective goods common to all residents. They are provided through various public actions, which may, however, be thwarted by the actions of developers trying to capture rent gains. What is at stake is whether consumers can effectively communicate their demands for the variety of amenities produced in land development. The results of this study suggest that further research should be aimed at providing information on the relationship between distribution of rent gains and the ability of various people to express their demands for suburban land features. In addition, attention should be given to certain products and costs which are not now relevant to the profit accounts of developers. More detailed knowledge of the effect of property rights on the size and incidence of costs (or benefits) associated with land transfer and conversion is critical if people are to control the type of communities that are being created by the transformation of rural land to urban uses.

A criticism commonly leveled against the typical suburban product is that there is inadequate open space and landscape amenity is destroyed. Is this the view merely of a few designers and planners with superior tastes that most lot and home buyers do not share? Or is there reason to believe that a general desire for more open space and similar amenities exists but has no means of expression?[1] Some city planners have tried to meet these criteria, but they have responded to demands expressed not through the market but through the political process. Many such plans seem to be gathering dust and implementation has been slow, and one might be in-

---

[1] For a general discussion of this issue, see the author's "Quality of the Environment and Man: Some Thoughts on Economic Institutions," *Journal of Soil and Water Conservation*, May-June 1966, pp. 89-91, and "Nonmarket Values and Efficiency of Public Investments in Water Resources," *American Economic Review*, May 1967, pp. 158-68.

clined to say that this is to be expected because they are bucking strong economic forces.

Two features of the problem need careful examination. Ownership of land does not usually follow landscape features very well. For example, one developer may find that the property he owns consists predominantly of a sharply sloped stream valley. It may well be that if the stream valley is left more or less intact the value of the total area proximate to it would be greater than if every square foot were developed for lots. But the owner of the stream land has an asset which he cannot produce and charge for unless he owns the surrounding land also, and he will fight any zoning which asks him to forego economic rent gains while his neighbors get the benefits. External benefits such as this are, of course, a great force arguing for enlargement of the scale of the developing firm so that the benefits become internalized and relevant to its enterprise. Yet there may be substantial barriers and objections to one firm gaining the necessary control, and they need further exploration.

An alternative to a single firm is some sort of a contractual arrangement between owners to share in the revenue of the whole area when developed. For example, a group of owners could agree to share in the appreciated value of their properties, regardless of which parcel sold first or where the various uses of their combined land were located.[2]

Another example of how distribution can affect the use made of land is the relationship between high and low density uses such as the residential and the commercial. The greatest land value appreciation is enjoyed by shopping center sites following growth of the surrounding housing. Typically this gain has accrued in a haphazard fashion to the lucky or influential owner of the particular tract favored by planners and the city council after the area's housing has been built up. I would hypothesize that this phenomenon has attracted the owners of the new towns of Reston, Virginia and Columbia, Maryland, who, while creating housing for a large area, have also been able to capture the rent gains for the shopping center and industry. These have apparently given them some slack within which they can create open space and other environmental amenities for the whole community. The spreading of certain overhead items may be important. Here it can be seen that the distribution of gains can affect the kinds of products made available. If these gains go to the individual owner of a shopping center site, they will never be used to create open space.

What are the implications of this for marketing rules? Perhaps it should be made easier for developers to put together large tracts, even by granting them powers of eminent domain. Otherwise, the power of the owners of the last few parcels to be acquired by a new town developer might prevent

---

[2] Such a proposal has been spelled out in terms of a real estate syndicate by Wallace-McHarg Associates, *Plan For The Valleys* (Philadelphia: Green Spring and Worthington Valley Planning Council, 1964).

such tracts from being assembled. Developers of large integrated tracts might be given the same power to issue tax free bonds for utilities that is now given to municipalities.

Development integration could also be provided by some type of special district as a form of local government, thus promoting the internalization of various externalities without resorting to a single large private development firm.[3] It would also provide a broader procedure for the expression of consumer demands and it might help to reduce the risks and unpredictable elements now present in the land conversion process.

The above are examples of cases where amenities are not produced because, even though consumers might be willing to pay for them, there is no way that the developers can charge for them, and the rent gains might be captured by an inappropriate party. In related situations attempts to capture the rent gains by particular parties create costs for others. For example, there is often great individual gain available to the owner who can build an apartment house in the middle of a planned open space or low density area, even though it may create congestion or expense for others, necessitate costly public services, or reduce the total value of the larger area of which it is a part. This promise of great gain to somebody exerts tremendous pressure against maintenance of a plan of development, which is often more than even the most angelic legislative body could be expected to resist. It may explain a great many unimplemented plans; as long as there are great premiums for breaking them, they will be in peril. One lesson for public policy design is that it should attempt to make private incentive consistent with the aims of public economic policy, not contrary to them.

This, of course, raises the question of whether it is desirable that an economy should be so structured that large rent gains were created to be fought over. It is quite possible to use market bargaining to ration a scarce commodity without allowing all rents to accrue to the owners. Our society has said that property rights do not include the right to the fruits of contrived monopoly. Should the profits (we call them "rents") accrue to those who have title to natural monopolies? If we forbid artificially contrived monopoly, should our attitude toward natural monopoly be any different? *The real importance of the distribution of rent gains is not one of equity alone, but of what kinds of products result.*

To summarize, if people wish to change the suburban products they are getting, one avenue is, of course, the traditional procedure of the market, where the prices of desired products are bid up. But in some cases this may not present a satisfactory range of alternatives to consumers, and the market rules and property rights must be modified to change the distribution of costs and benefits. This can be done by (1) internalizing some of the bene-

---

[3] Such a proposal has been made in detail by Marion Clawson, "Suburban Development Districts," *Journal of American Institute of Planners*, May 1960.

fits or costs, through the encouragement of larger firms or by facilitation of contractual arrangements which spread the gains by coordinating several owners or make certain effects costs to be accounted for, or (2) removing some of the (rent) gain for all parties.

Appeal to consumer sovereignty has been one of the great devices to end debate over whether the economy is actually performing as people want. It is not uncommon to hear the statement that they must prefer the present suburban products because that is what they in fact purchase. It is true that consumers can make land developers take certain things into account by buying or not buying; however, certain things cannot be demanded this way because there exists a set of property or distributional rules which influences the kinds of costs and benefits the developer takes into account and thus the range of products offered. There is nothing in the results of land market bargaining which would indicate people's choice as to what items should be taken into account. The choice of rules can never be unilateral but must always be made by the parties in common through a political process.

In addition, the question of the distribution of rent gain or appreciation is not merely a question of income distribution at a point in time. It influences the range of suburban products made available. It has its primary impact as developers try to influence public policies as expressed in land use plans and zoning in order to obtain the rent gain. Therefore, the issue is not just whether land is too highly priced and income is distributed to speculators and developers at the expense of home buyers, but whether people can express themselves on what kind of suburban communities they want to live in.

High priority should be given to research which will show the relationship between the range of suburban products available and the distribution of property rights, profits, and rents, so that people can make choices as true sovereigns over their economy. If consumer sovereignty is to have any reality, it must mean that people choose the game they want to play and the broad range of opportunities that should be available as well as how to play a given game to their advantage.

# APPENDIX

Appendix Table A-1. Lot Price Appreciation Above Farm Land and Improvement Costs, FHA Data, 1964

| City[a] and state | Price per lot[b] | Farm value per acre[c] | Lots per acre[d] | Farm value per lot (2) ÷ (3) | Farm value plus improvement cost (4) + 2,435 | Appreciation (1) − (5) | Appreciation over costs (6) ÷ (5) · 100 | Appreciation over farm value (6) ÷ (4) · 100 |
|---|---|---|---|---|---|---|---|---|
| | (1) | (2) | (3) | (4) | (5) | (6) | (7) | (8) |
| Albany, N.Y. | $2,425 | $165 | 2.57 | $ 64 | $2,499 | $ −74 | n.a.% | n.a.% |
| Albuquerque, N.M. | 2,675 | 30 | 2.37 | 13 | 2,448 | 227 | 9 | 1,793 |
| Atlanta, Ga. | 2,620 | 127 | 2.33 | 54 | 2,489 | 131 | 5 | 240 |
| Baltimore, Md. | 2,321 | 365 | 4.07 | 90 | 2,525 | −204 | n.a. | n.a. |
| Birmingham, Ala. | 2,496 | 116 | 2.33 | 50 | 2,485 | 11 | 0 | 23 |
| Boise, Idaho | 1,810 | 125 | 3.43 | 36 | 2,471 | −661 | n.a. | n.a. |
| Buffalo, N.Y. | 2,783 | 165 | 3.00 | 55 | 2,490 | 293 | 12 | 533 |
| Chicago, Ill. | 3,453 | 348 | 3.73 | 93 | 2,528 | 925 | 37 | 992 |
| Cincinnati, Ohio | 3,434 | 282 | 2.47 | 114 | 2,549 | 885 | 35 | 774 |
| Columbia, S.C. | 2,155 | 163 | 2.47 | 66 | 2,501 | −346 | n.a. | n.a. |
| Columbus, Ohio | 3,001 | 282 | 3.27 | 86 | 2,521 | 480 | 19 | 556 |
| Dallas, Texas | 2,210 | 108 | 3.50 | 31 | 2,466 | −256 | n.a. | n.a. |
| Denver, Colo. | 2,928 | 64 | 3.00 | 21 | 2,456 | 472 | 19 | 2,211 |
| Detroit, Mich. | 2,801 | 218 | 3.83 | 57 | 2,492 | 309 | 12 | 544 |
| Fort Worth, Tex. | 1,790 | 108 | 2.93 | 37 | 2,472 | −682 | n.a. | n.a. |
| Grand Rapids, Mich. | 2,387 | 218 | 2.57 | 85 | 2,520 | −133 | n.a. | n.a. |
| Greensboro, N.C. | 2,376 | 234 | 1.90 | 123 | 2,558 | −182 | n.a. | n.a. |
| Hartford, Conn. | 2,981 | 500 | 1.87 | 268 | 2,703 | 278 | 10 | 104 |
| Houston, Tex. | 2,420 | 108 | 3.80 | 28 | 2,463 | −43 | n.a. | n.a. |
| Indianapolis, Ind. | 2,538 | 293 | 2.77 | 106 | 2,541 | −3 | 0 | 0 |
| Jackson, Miss. | 2,007 | 135 | 2.60 | 52 | 2,487 | −480 | n.a. | n.a. |
| Jacksonville, Fla. | 2,412 | 307 | 2.43 | 126 | 2,561 | −149 | n.a. | n.a. |
| Kansas City, Mo. | 2,618 | 139 | 3.13 | 44 | 2,479 | 139 | 6 | 313 |
| Knoxville, Tenn. | 2,178 | 165 | 2.03 | 81 | 2,516 | −338 | n.a. | n.a. |
| Little Rock, Ark. | 2,826 | 152 | 2.83 | 54 | 2,489 | 337 | 14 | 629 |
| Los Angeles, Calif. | 5,178 | 460 | 3.80 | 121 | 2,556 | 2,622 | 103 | 2,166 |

| | | | | | | | |
|---|---|---|---|---|---|---|---|
| Louisville, Ky. | 2,632 | 171 | 3.13 | 55 | 2,490 | 142 | 6 | 261 |
| Lubbock, Tex. | 2,066 | 108 | 3.30 | 33 | 2,468 | −402 | n.a. | n.a. |
| Memphis, Tenn. | 2,504 | 165 | 2.80 | 59 | 2,494 | 10 | 0 | 17 |
| Miami, Fla. | 3,341 | 307 | 3.27 | 94 | 2,529 | 812 | 32 | 864 |
| Milwaukee, Wis. | 4,025 | 143 | 3.50 | 41 | 2,476 | 1,549 | 63 | 3,792 |
| Minneapolis, Minn. | 2,812 | 168 | 2.60 | 65 | 2,500 | 312 | 12 | 483 |
| New Orleans, La. | 4,211 | 213 | 4.03 | 53 | 2,488 | 1,723 | 69 | 3,263 |
| Oklahoma City, Okla. | 2,204 | 109 | 3.40 | 32 | 2,467 | −263 | n.a. | n.a. |
| Omaha, Nebr. | 2,961 | 104 | 2.97 | 35 | 2,470 | 491 | 20 | 1,400 |
| Philadelphia, Pa. | 2,757 | 222 | 5.87 | 38 | 2,473 | 284 | 11 | 751 |
| Phoenix, Ariz. | 2,587 | 60 | 3.30 | 18 | 2,453 | 134 | 5 | 736 |
| Pittsburgh, Pa. | 3,424 | 222 | 2.87 | 77 | 2,512 | 912 | 36 | 1,177 |
| Portland, Oreg. | 2,466 | 99 | 3.03 | 33 | 2,468 | −2 | 0 | 0 |
| Providence, R.I. | 1,915 | 417 | 2.33 | 179 | 2,614 | −699 | n.a. | n.a. |
| Reno, Nevada | 3,565 | 36 | 3.87 | 9 | 2,444 | 1,121 | 46 | 12,037 |
| Richmond, Va. | 2,890 | 164 | 2.73 | 60 | 2,495 | 395 | 16 | 658 |
| Sacramento, Calif. | 4,051 | 460 | 3.40 | 135 | 2,570 | 1,481 | 58 | 1,094 |
| Salt Lake City, Utah | 2,543 | 68 | 3.47 | 20 | 2,455 | 88 | 4 | 451 |
| San Antonio, Texas | 2,192 | 108 | 3.33 | 32 | 2,467 | −275 | n.a. | n.a. |
| San Diego, Calif. | 5,138 | 460 | 3.50 | 131 | 2,566 | 2,572 | 100 | 1,957 |
| San Francisco, Calif. | 4,451 | 460 | 3.93 | 117 | 2,552 | 1,899 | 74 | 1,624 |
| Santa Ana, Calif. | 6,429 | 460 | 3.87 | 119 | 2,554 | 3,875 | 152 | 3,257 |
| Seattle, Wash. | 3,076 | 145 | 2.80 | 52 | 2,487 | 589 | 24 | 1,138 |
| Shreveport, La. | 2,713 | 213 | 2.50 | 85 | 2,520 | 193 | 8 | 226 |
| Spokane, Wash. | 2,135 | 145 | 3.20 | 45 | 2,480 | −345 | n.a. | n.a. |
| St. Louis, Mo. | 3,264 | 139 | 3.10 | 45 | 2,480 | 784 | 32 | 1,749 |
| Tampa, Fla. | 2,692 | 307 | 3.37 | 91 | 2,526 | 166 | 7 | 182 |
| Tulsa, Okla. | 2,674 | 109 | 3.43 | 32 | 2,467 | 207 | 8 | 653 |
| Washington, D.C. | 3,925 | 365 | 2.50 | 146 | 2,581 | 1,344 | 52 | 921 |
| Wilmington, Del. | 2,905 | 295 | 2.90 | 102 | 2,537 | 368 | 15 | 362 |
| Average | | | | | | | | 892 |

n.a. = Not applicable.

a Housing area boundaries conform to SMSA boundaries when area is so designated, otherwise to county boundaries.

b Federal Housing Administration, Division of Research and Statistics, Table 8-M.

c State average farm value from U.S. Department of Agriculture, *Farm Real Estate Market Developments*, CD-66, pp. 13-14.

d From Table 6, average for 1964-1965.

Appendix Table A–2.   Lot Price Appreciation above Farm Land and Improvement Costs, NAHB Data, 1960

| State and city | Price per lot[a] | Farm value per lot[b] | Farm value plus improvement cost (2) + 2,435 | Appreciation (1) − (3) | Per cent appreciation over costs (4)÷(3)·100 | Per cent appreciation over farm value (4)÷(2)·100 |
|---|---|---|---|---|---|---|
| | (1) | (2) | (3) | (4) | (5) | (6) |
| *Alabama* | | | | | | |
| Birmingham | $3,583 | $35 | $2,470 | $1,113 | 45% | 3,180% |
| Decatur | 2,500 | 35 | 2,470 | 30 | 1 | 86 |
| Dothan | 1,650 | 35 | 2,470 | −820 | −33 | −2,343 |
| Huntsville | 2,682 | 35 | 2,470 | 212 | 9 | 606 |
| Mobile | 3,050 | 35 | 2,470 | 580 | 23 | 1,657 |
| Montgomery | 3,240 | 35 | 2,470 | 770 | 31 | 2,200 |
| Tuscaloosa | 2,850 | 35 | 2,470 | 380 | 15 | 1,086 |
| *Arizona* | | | | | | |
| Phoenix | 2,767 | 18 | 2,453 | 314 | 13 | 1,698 |
| Tucson | 2,727 | 18 | 2,453 | 274 | 11 | 1,482 |
| *Arkansas* | | | | | | |
| Fort Smith | 1,700 | 43 | 2,478 | −778 | −31 | −1,791 |
| Little Rock | 2,769 | 43 | 2,478 | 291 | 12 | 668 |
| Pine Bluff | 2,560 | 43 | 2,478 | 82 | 3 | 188 |
| *California* | | | | | | |
| Berkeley | 4,555 | 138 | 2,573 | 1,982 | 77 | 1,431 |
| Eureka | 2,500 | 138 | 2,573 | −73 | −3 | −53 |
| Fresno | 1,620 | 138 | 2,573 | −953 | −37 | −689 |
| Los Angeles | 5,110 | 138 | 2,573 | 2,537 | 99 | 1,832 |
| Modesto | 3,006 | 138 | 2,573 | 433 | 17 | 312 |
| Sacramento | 3,191 | 138 | 2,573 | 618 | 24 | 446 |
| San Bernardino | 3,675 | 138 | 2,573 | 1,102 | 43 | 796 |
| San Diego | 5,396 | 138 | 2,573 | 2,823 | 110 | 2,039 |
| San Francisco | 6,303 | 138 | 2,573 | 3,730 | 145 | 2,694 |
| San Mateo | 10,972 | 138 | 2,573 | 8,399 | 326 | 6,066 |
| Santa Barbara | 4,600 | 138 | 2,573 | 2,027 | 79 | 1,464 |
| Santa Clara | 5,269 | 138 | 2,573 | 2,696 | 105 | 1,947 |
| *Colorado* | | | | | | |
| Colorado Springs | 2,559 | 21 | 2,456 | 103 | 4 | 497 |
| Denver | 4,106 | 21 | 2,456 | 1,650 | 67 | 7,946 |
| Pueblo | 1,850 | 21 | 2,456 | −606 | −25 | −2,917 |
| *Connecticut* | | | | | | |
| Hamden | 3,071 | 172 | 2,607 | 464 | 18 | 271 |
| New London | 2,286 | 172 | 2,607 | −321 | −12 | −187 |
| West Hartford | 4,098 | 172 | 2,607 | 1,491 | 57 | 869 |
| *Delaware* | | | | | | |
| Wilmington | 3,731 | 93 | 2,528 | 1,203 | 48 | 1,287 |
| *Washington, D.C.* | 4,353 | 112 | 2,547 | 1,806 | 71 | 1,620 |
| *Florida* | | | | | | |
| Daytona Beach | 2,775 | 83 | 2,518 | 257 | 10 | 307 |
| Gainesville | 2,332 | 83 | 2,518 | −186 | −7 | −223 |
| Jacksonville | 2,577 | 83 | 2,518 | 59 | 2 | 70 |
| Lakeland | 3,425 | 83 | 2,518 | 907 | 36 | 1,086 |
| Miami | 3,369 | 83 | 2,518 | 851 | 34 | 1,019 |
| Orlando | 2,453 | 83 | 2,518 | −65 | −3 | −78 |
| Pensacola | 2,290 | 83 | 2,518 | −228 | −9 | −274 |
| St. Petersburg | 2,783 | 83 | 2,518 | 265 | 11 | 317 |
| Sarasota | 2,186 | 83 | 2,518 | −332 | −13 | −398 |
| Tallahassee | 2,110 | 83 | 2,518 | −408 | −16 | −489 |
| Tampa | 2,650 | 83 | 2,518 | 132 | 5 | 158 |

Appendix Table A–2.    (*Continued*)

| State and city | Price per lot[a] | Farm value per lot[b] | Farm value plus improvement cost (2) + 2,435 | Appreciation (1) − (3) | Per cent appreciation over costs (4) ÷ (3) · 100 | Per cent appreciation over farm value (4) ÷ (2) · 100 |
|---|---|---|---|---|---|---|
| | (1) | (2) | (3) | (4) | (5) | (6) |
| *Georgia* | | | | | | |
| Albany | 3,360 | 38 | 2,473 | 887 | 36 | 2,329 |
| Athens | 2,500 | 38 | 2,473 | 27 | 1 | 71 |
| Atlanta | 3,642 | 38 | 2,473 | 1,169 | 47 | 3,070 |
| Augusta | 2,480 | 38 | 2,473 | 7 | 0 | 18 |
| Columbus | 1,982 | 38 | 2,473 | −491 | −20 | −1,290 |
| Macon | 3,139 | 38 | 2,473 | 666 | 27 | 1,749 |
| Rome | 2,071 | 38 | 2,473 | −402 | −16 | −1,056 |
| Savannah | 2,428 | 38 | 2,473 | −45 | −2 | −118 |
| *Idaho* | | | | | | |
| Boise | 1,667 | 43 | 2,478 | −811 | −33 | −1,883 |
| *Illinois* | | | | | | |
| Alton | 2,650 | 122 | 2,557 | 93 | 4 | 77 |
| Belleville | 2,329 | 122 | 2,557 | −228 | −9 | −187 |
| Bloomington | 4,500 | 122 | 2,557 | 1,943 | 76 | 1,599 |
| Champaign | 3,938 | 122 | 2,557 | 1,381 | 54 | 1,137 |
| Chicago | 4,770 | 122 | 2,557 | 2,213 | 87 | 1,821 |
| Danville | 2,500 | 122 | 2,557 | −57 | −2 | −47 |
| Decatur | 7,633 | 122 | 2,557 | 5,076 | 199 | 4,177 |
| Galesburg | 1,900 | 122 | 2,557 | −657 | −26 | −540 |
| Kankakee | 2,950 | 122 | 2,557 | 393 | 15 | 324 |
| Rock Island | 3,513 | 122 | 2,557 | 956 | 37 | 787 |
| Peoria | 4,329 | 122 | 2,557 | 1,772 | 69 | 1,458 |
| Rockford | 2,710 | 122 | 2,557 | 153 | 6 | 126 |
| Springfield | 4,500 | 122 | 2,557 | 1,943 | 76 | 1,599 |
| *Indiana* | | | | | | |
| Anderson | 2,130 | 102 | 2,537 | −407 | −16 | −400 |
| Bloomington | 2,400 | 102 | 2,537 | −137 | −5 | −134 |
| Evansville | 2,500 | 102 | 2,537 | −37 | −1 | −36 |
| Fort Wayne | 2,964 | 102 | 2,537 | 427 | 17 | 421 |
| Hammond | 2,933 | 102 | 2,537 | 396 | 16 | 390 |
| Indianapolis | 3,010 | 102 | 2,537 | 473 | 19 | 466 |
| Lafayette | 3,300 | 102 | 2,537 | 763 | 30 | 752 |
| Marion | 1,500 | 102 | 2,537 | −1,037 | −41 | −1,021 |
| Muncie | 2,600 | 102 | 2,537 | 63 | 3 | 62 |
| New Albany | 2,100 | 102 | 2,537 | −437 | −17 | −430 |
| Richmond | 2,720 | 102 | 2,537 | 183 | 7 | 181 |
| South Bend | 2,706 | 102 | 2,537 | 169 | 7 | 167 |
| *Iowa* | | | | | | |
| Ames | 3,600 | 99 | 2,534 | 1,066 | 42 | 1,079 |
| Cedar Rapids | 3,877 | 99 | 2,534 | 1,343 | 53 | 1,359 |
| Council Bluffs | 2,450 | 99 | 2,534 | −84 | −3 | −85 |
| Davenport | 3,117 | 99 | 2,534 | 583 | 23 | 590 |
| Des Moines | 3,158 | 99 | 2,534 | 624 | 25 | 631 |
| Fort Dodge | 1,500 | 99 | 2,534 | −1,034 | −41 | −1,046 |
| Iowa City | 2,675 | 99 | 2,534 | 141 | 6 | 143 |
| Sioux City | 2,000 | 99 | 2,534 | −534 | −21 | −540 |
| *Kansas* | | | | | | |
| Hutchinson | 3,842 | 39 | 2,474 | 1,368 | 55 | 3,522 |
| Salina | 2,025 | 39 | 2,474 | −449 | −18 | −1,155 |
| Topeka | 3,423 | 39 | 2,474 | 949 | 38 | 2,443 |
| Wichita | 1,881 | 39 | 2,474 | −593 | −24 | −1,526 |
| *Kentucky* | | | | | | |
| Covington | 2,333 | 53 | 2,488 | −155 | −6 | −294 |
| Lexington | 3,344 | 53 | 2,488 | 856 | 34 | 1,625 |
| Louisville | 3,507 | 53 | 2,488 | 1,019 | 41 | 1,934 |
| Owensboro | 2,948 | 53 | 2,488 | 460 | 19 | 874 |
| Paducah | 2,467 | 53 | 2,488 | −21 | −1 | −39 |

Appendix Table A–2.  (*Continued*)

| State and city | Price per lot[a] (1) | Farm value per lot[b] (2) | Farm value plus improvement cost (2) + 2,435 (3) | Appreciation (1) − (3) (4) | Per cent appreciation over costs (4) ÷ (3) · 100 (5) | Per cent appreciation over farm value (4) ÷ (2) · 100 (6) |
|---|---|---|---|---|---|---|
| *Louisiana* | | | | | | |
| Alexandria | 3,133 | 67 | 2,502 | 631 | 25 | 949 |
| Baton Rouge | 3,210 | 67 | 2,502 | 708 | 28 | 1,065 |
| Lafayette | 2,733 | 67 | 2,502 | 231 | 9 | 348 |
| Lake Charles | 3,200 | 67 | 2,502 | 698 | 28 | 1,050 |
| New Orleans | 4,351 | 67 | 2,502 | 1,849 | 74 | 2,780 |
| Shreveport | 2,688 | 67 | 2,502 | 186 | 7 | 280 |
| *Maine* | | | | | | |
| Portland | 1,650 | 31 | 2,466 | −816 | −33 | −2,651 |
| *Maryland* | | | | | | |
| Baltimore | 2,895 | 112 | 2,547 | 348 | 14 | 312 |
| Cumberland | 1,940 | 112 | 2,547 | −607 | −24 | −544 |
| *Massachusetts* | | | | | | |
| Attleboro | 2,125 | 121 | 2,556 | −431 | −17 | −357 |
| Boston | 3,694 | 121 | 2,556 | 1,138 | 45 | 942 |
| Fall River | 2,250 | 121 | 2,556 | −306 | −12 | −253 |
| Lowell | 2,050 | 121 | 2,556 | −506 | −20 | −419 |
| Lawrence | 4,000 | 121 | 2,556 | 1,444 | 57 | 1,196 |
| Springfield | 2,307 | 121 | 2,556 | −249 | −10 | −206 |
| Worcester | 2,838 | 121 | 2,556 | 282 | 11 | 234 |
| *Michigan* | | | | | | |
| Ann Arbor | 3,700 | 75 | 2,510 | 1,190 | 47 | 1,595 |
| Battle Creek | 2,733 | 75 | 2,510 | 223 | 9 | 299 |
| Detroit | 4,155 | 75 | 2,510 | 1,645 | 66 | 2,205 |
| Flint | 2,157 | 75 | 2,510 | −353 | −14 | −473 |
| Grand Rapids | 2,913 | 75 | 2,510 | 403 | 16 | 541 |
| Jackson | 2,686 | 75 | 2,510 | 176 | 7 | 236 |
| Kalamazoo | 2,629 | 75 | 2,510 | 119 | 5 | 160 |
| Lansing | 3,161 | 75 | 2,510 | 651 | 26 | 873 |
| Midland | 3,000 | 75 | 2,510 | 490 | 20 | 657 |
| Muskegon | 2,225 | 75 | 2,510 | −285 | −11 | −381 |
| Port Huron | 1,233 | 75 | 2,510 | −1,277 | −51 | −1,711 |
| Saginaw | 2,700 | 75 | 2,510 | 190 | 8 | 255 |
| *Minnesota* | | | | | | |
| Duluth | 3,900 | 60 | 2,495 | 1,405 | 56 | 2,357 |
| Minneapolis | 3,080 | 60 | 2,495 | 585 | 23 | 982 |
| Rochester | 3,833 | 60 | 2,495 | 1,338 | 54 | 2,245 |
| St. Paul | 2,550 | 60 | 2,495 | 55 | 2 | 93 |
| *Mississippi* | | | | | | |
| Greenville | 1,830 | 41 | 2,476 | −646 | −26 | −1,570 |
| Hattiesburg | 1,933 | 41 | 2,476 | −543 | −22 | −1,320 |
| Jackson | 2,645 | 41 | 2,476 | 169 | 7 | 410 |
| *Missouri* | | | | | | |
| Columbia | 2,400 | 44 | 2,479 | −79 | −3 | −179 |
| Kansas City | 3,172 | 44 | 2,479 | 693 | 28 | 1,566 |
| Springfield | 2,383 | 44 | 2,479 | −96 | −4 | −218 |
| St. Joseph | 1,780 | 44 | 2,479 | −699 | −28 | −1,581 |
| St. Louis | 4,878 | 44 | 2,479 | 2,399 | 97 | 5,423 |
| *Montana* | | | | | | |
| Billings | 2,150 | 13 | 2,448 | −298 | −12 | −2,217 |
| Great Falls | 2,455 | 13 | 2,448 | 7 | 0 | 49 |
| *Nebraska* | | | | | | |
| Lincoln | 2,625 | 35 | 2,470 | 155 | 6 | 449 |
| Omaha | 3,071 | 35 | 2,470 | 601 | 24 | 1,737 |

Appendix Table A–2.    (*Continued*)

| State and city | Price per lot[a] | Farm value per lot[b] | Farm value plus improvement cost (2) + 2,435 | Appreciation (1) − (3) | Per cent appreciation over costs (4) ÷ (3) · 100 | Per cent appreciation over farm value (4) ÷ (2) · 100 |
|---|---|---|---|---|---|---|
| | (1) | (2) | (3) | (4) | (5) | (6) |
| *Nevada* | | | | | | |
| Las Vegas | 2,792 | 12 | 2,447 | 345 | 14 | 2,894 |
| Reno | 3,211 | 12 | 2,447 | 764 | 31 | 6,408 |
| *New Hampshire* | | | | | | |
| Nashua | 2,500 | 40 | 2,475 | 25 | 1 | 63 |
| Manchester | 1,867 | 40 | 2,475 | −608 | −25 | −1,520 |
| *New Jersey* | | | | | | |
| Atlantic City | 1,988 | 203 | 2,638 | −650 | −25 | −320 |
| Camden | 3,051 | 203 | 2,638 | 413 | 16 | 203 |
| Union | 6,635 | 203 | 2,638 | 3,997 | 152 | 1,968 |
| Edison | 4,371 | 203 | 2,638 | 1,733 | 66 | 853 |
| Trenton | 5,583 | 203 | 2,638 | 2,945 | 112 | 1,450 |
| *New Mexico* | | | | | | |
| Albuquerque | 2,805 | 9 | 2,444 | 361 | 15 | 3,908 |
| Las Cruces | 2,183 | 9 | 2,444 | −261 | −11 | −2,830 |
| Roswell | 2,460 | 9 | 2,444 | 16 | 1 | 171 |
| Santa Fe | 2,350 | 9 | 2,444 | −94 | −4 | −1,021 |
| *New York* | | | | | | |
| Albany | 2,771 | 56 | 2,491 | 280 | 11 | 502 |
| Buffalo | 2,929 | 56 | 2,491 | 438 | 18 | 786 |
| Elmira | 2,857 | 56 | 2,491 | 366 | 15 | 657 |
| Hempstead | 4,979 | 56 | 2,491 | 2,488 | 100 | 4,462 |
| Poughkeepsie | 2,133 | 56 | 2,491 | −358 | −14 | −642 |
| Rochester | 3,729 | 56 | 2,491 | 1,238 | 50 | 2,220 |
| Rome | 2,620 | 56 | 2,491 | 129 | 5 | 232 |
| Schenectady | 2,689 | 56 | 2,491 | 198 | 8 | 355 |
| Syracuse | 3,178 | 56 | 2,491 | 687 | 28 | 1,232 |
| *North Carolina* | | | | | | |
| Charlotte | 2,579 | 72 | 2,507 | 72 | 3 | 101 |
| Durham | 2,407 | 72 | 2,507 | −100 | −4 | −139 |
| Fayetteville | 1,750 | 72 | 2,507 | −757 | −30 | −1,058 |
| Greensboro | 3,296 | 72 | 2,507 | 789 | 31 | 1,104 |
| Raleigh | 2,990 | 72 | 2,507 | 483 | 19 | 676 |
| Rocky Mount | 2,750 | 72 | 2,507 | 243 | 10 | 340 |
| Winston-Salem | 2,649 | 72 | 2,507 | 142 | 6 | 199 |
| *North Dakota* | | | | | | |
| Fargo | 1,863 | 20 | 2,455 | −592 | −24 | −2,906 |
| *Ohio* | | | | | | |
| Akron | 4,027 | 95 | 2,530 | 1,497 | 59 | 1,569 |
| Canton | 2,524 | 95 | 2,530 | −6 | −0 | −7 |
| Cincinnati | 4,190 | 95 | 2,530 | 1,660 | 66 | 1,740 |
| Cleveland | 4,442 | 95 | 2,530 | 1,912 | 76 | 2,004 |
| Columbus | 3,560 | 95 | 2,530 | 1,030 | 41 | 1,079 |
| Dayton | 3,589 | 95 | 2,530 | 1,059 | 42 | 1,110 |
| Elyria | 2,692 | 95 | 2,530 | 162 | 6 | 169 |
| Lancaster | 2,600 | 95 | 2,530 | 70 | 3 | 73 |
| Lima | 3,083 | 95 | 2,530 | 553 | 22 | 579 |
| Mansfield | 2,930 | 95 | 2,530 | 400 | 16 | 419 |
| Sandusky | 3,067 | 95 | 2,530 | 537 | 21 | 563 |
| Springfield | 2,850 | 95 | 2,530 | 320 | 13 | 335 |
| Toledo | 3,462 | 95 | 2,530 | 932 | 37 | 977 |
| Youngstown | 2,620 | 95 | 2,530 | 90 | 4 | 94 |
| *Oklahoma* | | | | | | |
| Bartlesville | 2,240 | 33 | 2,468 | −228 | −9 | −690 |
| Enid | 2,200 | 33 | 2,468 | −268 | −11 | −810 |
| Lawton | 1,650 | 33 | 2,468 | −818 | −33 | −2,473 |
| Norman | 2,328 | 33 | 2,468 | −140 | −6 | −423 |
| Oklahoma City | 2,694 | 33 | 2,468 | 226 | 9 | 683 |
| Tulsa | 2,915 | 33 | 2,468 | 447 | 18 | 1,351 |

Appendix Table A–2. (*Continued*)

| State and city | Price per lot[a] | Farm value per lot[b] | Farm value plus improvement cost (2) + 2,435 | Appreciation (1) − (3) | Per cent appreciation over costs (4) ÷ (3) · 100 | Per cent appreciation over farm value (4) ÷ (2) · 100 |
|---|---|---|---|---|---|---|
| | (1) | (2) | (3) | (4) | (5) | (6) |
| *Oregon* | | | | | | |
| Eugene | 2,475 | 34 | 2,469 | 6 | 0 | 18 |
| Portland | 2,604 | 34 | 2,469 | 135 | 5 | 399 |
| Salem | 2,159 | 34 | 2,469 | −310 | −13 | −915 |
| *Pennsylvania* | | | | | | |
| Allentown | 2,475 | 72 | 2,507 | −32 | −1 | −45 |
| Altoona | 1,500 | 72 | 2,507 | −1,007 | −40 | −1,393 |
| Erie | 2,133 | 72 | 2,507 | −374 | −15 | −518 |
| Harrisburg | 2,402 | 72 | 2,507 | −105 | −4 | −146 |
| Johnstown | 2,350 | 72 | 2,507 | −157 | −6 | −218 |
| Lancaster | 2,040 | 72 | 2,507 | −467 | −19 | −646 |
| Philadelphia | 3,064 | 72 | 2,507 | 557 | 22 | 770 |
| Pittsburgh | 3,932 | 72 | 2,507 | 1,425 | 57 | 1,970 |
| Reading | 1,655 | 72 | 2,507 | −852 | −34 | −1,179 |
| Williamsport | 2,140 | 72 | 2,507 | −367 | −15 | −508 |
| York | 2,209 | 72 | 2,507 | −298 | −12 | −413 |
| *Rhode Island* | | | | | | |
| Providence | 1,745 | 146 | 2,581 | −836 | −32 | −573 |
| *South Carolina* | | | | | | |
| Charleston | 2,472 | 53 | 2,488 | −16 | −1 | −30 |
| Columbia | 2,383 | 53 | 2,488 | −105 | −4 | −199 |
| Greenville | 1,879 | 53 | 2,488 | −609 | −24 | −1,155 |
| Spartanburg | 1,881 | 53 | 2,488 | −607 | −24 | −1,151 |
| *South Dakota* | | | | | | |
| Rapid City | 2,275 | 20 | 2,455 | −180 | −7 | −916 |
| Sioux Falls | 2,533 | 20 | 2,455 | 78 | 3 | 400 |
| *Tennessee* | | | | | | |
| Chattanooga | 2,168 | 51 | 2,486 | −318 | −13 | −626 |
| Jackson | 1,500 | 51 | 2,486 | −986 | −40 | −1,942 |
| Johnson City | 2,933 | 51 | 2,486 | 447 | 18 | 881 |
| Knoxville | 2,209 | 51 | 2,486 | −277 | −11 | −545 |
| Memphis | 2,961 | 51 | 2,486 | 475 | 19 | 936 |
| Nashville | 2,215 | 51 | 2,486 | −271 | −11 | −533 |
| Oak Ridge | 1,500 | 51 | 2,486 | −986 | −40 | −1,942 |
| *Texas* | | | | | | |
| Abilene | 1,890 | 33 | 2,468 | −578 | −23 | −1,767 |
| Amarillo | 1,960 | 33 | 2,468 | −508 | −21 | −1,553 |
| Austin | 2,312 | 33 | 2,468 | −156 | −6 | −476 |
| Beaumont | 2,309 | 33 | 2,468 | −159 | −6 | −485 |
| Corpus Christi | 2,491 | 33 | 2,468 | 23 | 1 | 71 |
| Dallas | 3,201 | 33 | 2,468 | 733 | 30 | 2,243 |
| El Paso | 2,542 | 33 | 2,468 | 74 | 3 | 227 |
| Fort Worth | 2,798 | 33 | 2,468 | 330 | 13 | 1,010 |
| Houston | 3,425 | 33 | 2,468 | 957 | 39 | 2,928 |
| Longview | 2,025 | 33 | 2,468 | −443 | −18 | −1,354 |
| Lubbock | 3,400 | 33 | 2,468 | 932 | 38 | 2,852 |
| Midland | 3,000 | 33 | 2,468 | 532 | 22 | 1,628 |
| San Angelo | 1,492 | 33 | 2,468 | −976 | −40 | −2,984 |
| San Antonio | 2,295 | 33 | 2,468 | −173 | −7 | −528 |
| Texarkana | 1,875 | 33 | 2,468 | −593 | −24 | −1,813 |
| Tyler | 2,200 | 33 | 2,468 | −268 | −11 | −819 |
| Victoria | 1,822 | 33 | 2,468 | −646 | −26 | −1,975 |
| Waco | 1,592 | 33 | 2,468 | −876 | −35 | −2,679 |
| Wichita Falls | 2,583 | 33 | 2,468 | 115 | 5 | 353 |

Appendix Table A–2.    (*Continued*)

| State and city | Price per lot[a] (1) | Farm value per lot[b] (2) | Farm value plus improvement cost (2) + 2,435 (3) | Appreciation (1) − (3) (4) | Per cent appreciation over costs (4) ÷ (3) · 100 (5) | Per cent appreciation over farm value (4) ÷ (2) · 100 (6) |
|---|---|---|---|---|---|---|
| *Utah* | | | | | | |
| Ogden | 2,310 | 23 | 2,458 | −148 | −6 | −642 |
| Salt Lake City | 2,635 | 23 | 2,458 | 177 | 7 | 767 |
| *Vermont* | | | | | | |
| Burlington | 2,700 | 31 | 2,466 | 234 | 9 | 751 |
| *Virginia* | | | | | | |
| Charlottesville | 2,929 | 54 | 2,489 | 440 | 18 | 817 |
| Danville | 1,350 | 54 | 2,489 | −1,139 | −46 | −2,115 |
| Lynchburg | 1,767 | 54 | 2,489 | −722 | −29 | −1,341 |
| Newport News | 2,977 | 54 | 2,489 | 488 | 20 | 907 |
| Norfolk | 2,505 | 54 | 2,489 | 16 | 1 | 30 |
| Petersburg | 1,745 | 54 | 2,489 | −744 | −30 | −1,381 |
| Richmond | 2,750 | 54 | 2,489 | 261 | 10 | 485 |
| Roanoke | 1,825 | 54 | 2,489 | −664 | −27 | −1,233 |
| *Washington* | | | | | | |
| Seattle | 2,688 | 51 | 2,486 | 202 | 8 | 395 |
| Spokane | 2,644 | 51 | 2,486 | 158 | 6 | 309 |
| Tacoma | 2,147 | 51 | 2,486 | −339 | −14 | −663 |
| Yakima | 3,250 | 51 | 2,486 | 764 | 31 | 1,493 |
| *West Virginia* | | | | | | |
| Charleston | 3,675 | 29 | 2,464 | 1,211 | 49 | 4,199 |
| Huntington | 2,875 | 29 | 2,464 | 411 | 17 | 1,425 |
| Wheeling | 2,625. | 29 | 2,464 | 161 | 7 | 559 |
| *Wisconsin* | | | | | | |
| Appleton | 3,133 | 51 | 2,486 | 647 | 26 | 1,265 |
| Beloit | 1,654 | 51 | 2,486 | −832 | −33 | −1,627 |
| Green Bay | 1,783 | 51 | 2,486 | −703 | −28 | −1,375 |
| Madison | 6,250 | 51 | 2,486 | 3,764 | 151 | 7,358 |
| Milwaukee | 4,100 | 51 | 2,486 | 1,614 | 65 | 3,155 |
| Racine | 2,883 | 51 | 2,486 | 397 | 16 | 776 |
| *Wyoming* | | | | | | |
| Cheyenne | 2,100 | 8 | 2,443 | −343 | −14 | −4,059 |
| High | 10,972 | 203 | 2,638 | 8,399 | 326 | 7,946 |
| Low | 1,233 | 8 | 2,443 | −1,277 | −51 | −4,059 |
| Average | 2,857 | 70 | 2,505 | 351 | 14 | 399 |

[a] Lot price data from National Association of Home Builders, *Economic News Notes*, Special Report 65–8 (Washington).

[b] Farm value per acre as shown in Appendix Table A–7 divided by a constant 2.6 lots per acre. A slight improvement in this variable might be obtained by using the FHA average lot sizes shown in Table 6. However, since the data are not available for all the cities in this table, and FHA districts may not match exactly with the NAHB cities, a constant size was used.

Appendix Table A–3. Lot Price Appreciation above Farm Land and Improvement Costs, NAHB Data, 1964

| State and city | Price per lot[a] | Farm value per lot[b] | Farm value plus improvement cost (2) + 2,435 | Appreciation (1) − (3) | Per cent appreciation over costs (4) ÷ (3) · 100 | Per cent appreciation over farm value (4) ÷ (2) · 100 |
|---|---|---|---|---|---|---|
| | (1) | (2) | (3) | (4) | (5) | (6) |
| *Alabama* | | | | | | |
| Birmingham | $3,939 | $ 45 | $2,480 | $1,459 | 59% | 3,271% |
| Decatur | 3,400 | 45 | 2,480 | 920 | 37 | 2,063 |
| Dothan | 2,083 | 45 | 2,480 | −397 | −16 | −889 |
| Huntsville | 3,650 | 45 | 2,480 | 1,170 | 47 | 2,623 |
| Mobile | 3,750 | 45 | 2,480 | 1,270 | 51 | 2,847 |
| Montgomery | 3,708 | 45 | 2,480 | 1,228 | 50 | 2,753 |
| Tuscaloosa | 3,750 | 45 | 2,480 | 1,270 | 51 | 2,847 |
| *Arizona* | | | | | | |
| Phoenix | 4,657 | 23 | 2,458 | 2,199 | 89 | 9,529 |
| Tucson | 4,145 | 23 | 2,458 | 1,687 | 69 | 7,310 |
| *Arkansas* | | | | | | |
| Fort Smith | 1,758 | 58 | 2,493 | −735 | −29 | −1,258 |
| Little Rock | 3,859 | 58 | 2,493 | 1,366 | 55 | 2,336 |
| Pine Bluff | 3,260 | 58 | 2,493 | 767 | 31 | 1,311 |
| *California* | | | | | | |
| Berkeley | 6,943 | 177 | 2,612 | 4,331 | 166 | 2,448 |
| Eureka | 3,500 | 177 | 2,612 | 888 | 34 | 502 |
| Fresno | 2,480 | 177 | 2,612 | −132 | −5 | −75 |
| Los Angeles | 9,855 | 177 | 2,612 | 7,243 | 277 | 4,094 |
| Modesto | 4,009 | 177 | 2,612 | 1,397 | 53 | 790 |
| Sacramento | 4,485 | 177 | 2,612 | 1,873 | 72 | 1,059 |
| San Bernardino | 5,959 | 177 | 2,612 | 3,347 | 128 | 1,892 |
| San Diego | 7,839 | 177 | 2,612 | 5,227 | 200 | 2,954 |
| San Francisco | 10,648 | 177 | 2,612 | 8,036 | 308 | 4,542 |
| San Mateo | 15,469 | 177 | 2,612 | 12,857 | 492 | 7,267 |
| Santa Barbara | 11,269 | 177 | 2,612 | 8,657 | 331 | 4,893 |
| Santa Clara | 9,700 | 177 | 2,612 | 7,088 | 271 | 4,006 |
| *Colorado* | | | | | | |
| Colorado Springs | 3,338 | 25 | 2,460 | 878 | 36 | 3,568 |
| Denver | 5,094 | 25 | 2,460 | 2,634 | 107 | 10,702 |
| Pueblo | 2,040 | 25 | 2,460 | −420 | −17 | −1,705 |
| *Connecticut* | | | | | | |
| Hamden | 5,035 | 192 | 2,627 | 2,408 | 92 | 1,252 |
| New London | 6,413 | 192 | 2,627 | 3,786 | 144 | 1,969 |
| West Hartford | 5,954 | 192 | 2,627 | 3,327 | 127 | 1,730 |
| *Delaware* | | | | | | |
| Wilmington | 5,833 | 113 | 2,548 | 3,285 | 129 | 2,895 |
| *Washington, D.C.* | 7,035 | 140 | 2,575 | 4,460 | 173 | 3,177 |
| *Florida* | | | | | | |
| Daytona Beach | 4,050 | 118 | 2,553 | 1,497 | 59 | 1,268 |
| Gainesville | 2,910 | 118 | 2,553 | 357 | 14 | 302 |
| Jacksonville | 3,213 | 118 | 2,553 | 660 | 26 | 559 |
| Lakeland | 3,498 | 118 | 2,553 | 945 | 37 | 800 |
| Miami | 3,968 | 118 | 2,553 | 1,415 | 55 | 1,198 |
| Orlando | 3,389 | 118 | 2,553 | 836 | 33 | 708 |
| Pensacola | 3,075 | 118 | 2,553 | 522 | 20 | 442 |
| St. Petersburg | 3,360 | 118 | 2,553 | 807 | 32 | 683 |
| Sarasota | 3,933 | 118 | 2,553 | 1,380 | 54 | 1,169 |
| Tallahassee | 2,679 | 118 | 2,553 | 126 | 5 | 107 |
| Tampa | 3,259 | 118 | 2,553 | 706 | 28 | 598 |

Appendix Table A–3. *(Continued)*

| State and city | Price per lot[a] | Farm value per lot[b] | Farm value plus improvement cost (2) + 2,435 | Appreciation (1) − (3) | Per cent appreciation over costs (4) ÷ (3) · 100 | Per cent appreciation over farm value (4) ÷ (2) · 100 |
|---|---|---|---|---|---|---|
| | (1) | (2) | (3) | (4) | (5) | (6) |
| *Georgia* | | | | | | |
| Albany | $4,280 | $ 49 | $2,484 | $1,796 | 72% | 3,677% |
| Athens | 3,088 | 49 | 2,484 | 604 | 24 | 1,237 |
| Atlanta | 5,027 | 49 | 2,484 | 2,543 | 102 | 5,206 |
| Augusta | 3,596 | 49 | 2,484 | 1,112 | 45 | 2,277 |
| Columbus | 2,853 | 49 | 2,484 | 369 | 15 | 756 |
| Macon | 4,464 | 49 | 2,484 | 1,980 | 80 | 4,054 |
| Rome | 2,614 | 49 | 2,484 | 130 | 5 | 266 |
| Savannah | 3,510 | 49 | 2,484 | 1,026 | 41 | 2,101 |
| *Idaho* | | | | | | |
| Boise | 2,183 | 48 | 2,483 | −300 | −12 | −624 |
| *Illinois* | | | | | | |
| Alton | 3,230 | 134 | 2,569 | 661 | 26 | 494 |
| Belleville | 2,743 | 134 | 2,569 | 174 | 7 | 130 |
| Bloomington | 4,250 | 134 | 2,569 | 1,681 | 65 | 1,256 |
| Champaign | 5,142 | 134 | 2,569 | 2,573 | 100 | 1,922 |
| Chicago | 6,228 | 134 | 2,569 | 3,659 | 142 | 2,734 |
| Danville | 4,200 | 134 | 2,569 | 1,631 | 63 | 1,219 |
| Decatur | 9,548 | 134 | 2,569 | 6,979 | 272 | 5,214 |
| Galesburg | 2,500 | 134 | 2,569 | −69 | −3 | −51 |
| Kankakee | 3,075 | 134 | 2,569 | 506 | 20 | 378 |
| Rock Island | 5,300 | 134 | 2,569 | 2,731 | 106 | 2,041 |
| Peoria | 5,533 | 134 | 2,569 | 2,964 | 115 | 2,215 |
| Rockford | 3,150 | 134 | 2,569 | 581 | 23 | 434 |
| Springfield | 3,300 | 134 | 2,569 | 731 | 28 | 546 |
| *Indiana* | | | | | | |
| Anderson | 2,938 | 113 | 2,548 | 390 | 15 | 346 |
| Bloomington | 3,836 | 113 | 2,548 | 1,288 | 51 | 1,143 |
| Evansville | 3,533 | 113 | 2,548 | 985 | 39 | 874 |
| Fort Wayne | 3,652 | 113 | 2,548 | 1,104 | 43 | 980 |
| Hammond | 3,288 | 113 | 2,548 | 740 | 29 | 657 |
| Indianapolis | 5,055 | 113 | 2,548 | 2,507 | 98 | 2,225 |
| Lafayette | 4,292 | 113 | 2,548 | 1,744 | 68 | 1,548 |
| Marion | 3,250 | 113 | 2,548 | 702 | 28 | 623 |
| Muncie | 2,125 | 113 | 2,548 | −423 | −17 | −375 |
| New Albany | 3,170 | 113 | 2,548 | 622 | 24 | 552 |
| Richmond | 3,017 | 113 | 2,548 | 469 | 18 | 416 |
| South Bend | 3,056 | 113 | 2,548 | 508 | 20 | 451 |
| *Iowa* | | | | | | |
| Ames | 3,886 | 102 | 2,537 | 1,349 | 53 | 1,324 |
| Cedar Rapids | 3,918 | 102 | 2,537 | 1,381 | 54 | 1,355 |
| Council Bluffs | 2,814 | 102 | 2,537 | 277 | 11 | 272 |
| Davenport | 3,529 | 102 | 2,537 | 992 | 39 | 973 |
| Des Moines | 3,774 | 102 | 2,537 | 1,237 | 49 | 1,214 |
| Fort Dodge | 3,500 | 102 | 2,537 | 963 | 38 | 945 |
| Iowa City | 3,900 | 102 | 2,537 | 1,363 | 54 | 1,337 |
| Sioux City | 1,850 | 102 | 2,537 | −687 | −27 | −674 |
| *Kansas* | | | | | | |
| Hutchinson | 4,907 | 44 | 2,479 | 2,428 | 98 | 5,538 |
| Salina | 2,200 | 44 | 2,479 | −279 | −11 | −636 |
| Topeka | 4,086 | 44 | 2,479 | 1,607 | 65 | 3,665 |
| Wichita | 3,105 | 44 | 2,479 | 626 | 25 | 1,428 |
| *Kentucky* | | | | | | |
| Covington | 3,766 | 66 | 2,501 | 1,265 | 51 | 1,924 |
| Lexington | 4,471 | 66 | 2,501 | 1,970 | 79 | 2,996 |
| Louisville | 4,401 | 66 | 2,501 | 1,900 | 76 | 2,889 |
| Owensboro | 3,121 | 66 | 2,501 | 620 | 25 | 943 |
| Paducah | 3,200 | 66 | 2,501 | 699 | 28 | 1,063 |

Appendix Table A–3.   (*Continued*)

| State and city | Price per lot[a] | Farm value per lot[b] | Farm value plus improvement cost (2) + 2,435 | Appreciation (1) − (3) | Per cent appreciation over costs (4) ÷ (3) · 100 | Per cent appreciation over farm value (4) ÷ (2) · 100 |
|---|---|---|---|---|---|---|
| | (1) | (2) | (3) | (4) | (5) | (6) |
| *Louisiana* | | | | | | |
| Alexandria | $4,113 | $ 82 | $2,517 | $1,596 | 63% | 1,948% |
| Baton Rouge | 3,960 | 82 | 2,517 | 1,443 | 57 | 1,762 |
| Lafayette | 3,733 | 82 | 2,517 | 1,216 | 48 | 1,484 |
| Lake Charles | 4,500 | 82 | 2,517 | 1,983 | 79 | 2,421 |
| New Orleans | 5,652 | 82 | 2,517 | 3,135 | 125 | 3,827 |
| Shreveport | 2,959 | 82 | 2,517 | 442 | 18 | 540 |
| *Maine* | | | | | | |
| Portland | 2,783 | 36 | 2,471 | 312 | 13 | 873 |
| *Maryland* | | | | | | |
| Baltimore | 4,342 | 140 | 2,575 | 1,767 | 69 | 1,258 |
| Cumberland | 2,650 | 140 | 2,575 | 75 | 3 | 53 |
| *Massachusetts* | | | | | | |
| Attleboro | 3,000 | 134 | 2,569 | 431 | 17 | 321 |
| Boston | 5,724 | 134 | 2,569 | 3,155 | 123 | 2,350 |
| Fall River | 2,650 | 134 | 2,569 | 81 | 3 | 60 |
| Lowell | 3,344 | 134 | 2,569 | 775 | 30 | 577 |
| Lawrence | 4,750 | 134 | 2,569 | 2,181 | 85 | 1,625 |
| Springfield | 3,123 | 134 | 2,569 | 554 | 22 | 413 |
| Worcester | 3,446 | 134 | 2,569 | 877 | 34 | 653 |
| *Michigan* | | | | | | |
| Ann Arbor | 4,563 | 84 | 2,519 | 2,044 | 81 | 2,438 |
| Battle Creek | 3,057 | 84 | 2,519 | 538 | 21 | 642 |
| Detroit | 5,765 | 84 | 2,519 | 3,246 | 129 | 3,872 |
| Flint | 2,944 | 84 | 2,519 | 425 | 17 | 507 |
| Grand Rapids | 4,328 | 84 | 2,519 | 1,809 | 72 | 2,158 |
| Jackson | 3,525 | 84 | 2,519 | 1,006 | 40 | 1,200 |
| Kalamazoo | 3,677 | 84 | 2,519 | 1,158 | 46 | 1,381 |
| Lansing | 4,300 | 84 | 2,519 | 1,781 | 71 | 2,124 |
| Midland | 3,800 | 84 | 2,519 | 1,281 | 51 | 1,528 |
| Muskegon | 2,138 | 84 | 2,519 | −381 | −15 | −454 |
| Port Huron | 1,633 | 84 | 2,519 | −886 | −35 | −1,057 |
| Saginaw | 3,113 | 84 | 2,519 | 594 | 24 | 709 |
| *Minnesota* | | | | | | |
| Duluth | 4,263 | 65 | 2,500 | 1,763 | 71 | 2,729 |
| Minneapolis | 4,296 | 65 | 2,500 | 1,796 | 72 | 2,780 |
| Rochester | 9,225 | 65 | 2,500 | 6,725 | 269 | 10,408 |
| St. Paul | 3,738 | 65 | 2,500 | 1,238 | 50 | 1,917 |
| *Mississippi* | | | | | | |
| Greenville | 2,425 | 52 | 2,487 | −62 | −2 | −119 |
| Hattiesburg | 2,617 | 52 | 2,487 | 130 | 5 | 251 |
| Jackson | 3,428 | 52 | 2,487 | 941 | 38 | 1,812 |
| *Missouri* | | | | | | |
| Columbia | 3,167 | 53 | 2,488 | 679 | 27 | 1,269 |
| Kansas City | 4,398 | 53 | 2,488 | 1,910 | 77 | 3,572 |
| Springfield | 2,970 | 53 | 2,488 | 482 | 19 | 901 |
| St. Joseph | 2,440 | 53 | 2,488 | −48 | −2 | −91 |
| St. Louis | 6,376 | 53 | 2,488 | 3,888 | 156 | 7,272 |
| *Montana* | | | | | | |
| Billings | 2,150 | 16 | 2,451 | −301 | −12 | −1,864 |
| Great Falls | 3,100 | 16 | 2,451 | 649 | 26 | 4,017 |
| *Nebraska* | | | | | | |
| Lincoln | 3,815 | 40 | 2,475 | 1,340 | 54 | 3,350 |
| Omaha | 3,751 | 40 | 2,475 | 1,276 | 52 | 3,190 |

Appendix Table A–3.  (*Continued*)

| State and city | Price per lot[a] | Farm value per lot[b] | Farm value plus improvement cost (2) + 2,435 | Appreciation (1) − (3) | Per cent appreciation over costs (4) ÷ (3) · 100 | Per cent appreciation over farm value (4) ÷ (2) · 100 |
|---|---|---|---|---|---|---|
| | (1) | (2) | (3) | (4) | (5) | (6) |
| *Nevada* | | | | | | |
| Las Vegas | $4,712 | $ 14 | $2,449 | $2,263 | 92% | 16,345% |
| Reno | 4,700 | 14 | 2,449 | 2,251 | 92 | 16,258 |
| *New Hampshire* | | | | | | |
| Nashua | 3,500 | 45 | 2,480 | 1,020 | 41 | 2,247 |
| Manchester | 2,350 | 45 | 2,480 | −130 | −5 | −287 |
| *New Jersey* | | | | | | |
| Atlantic City | 3,163 | 231 | 2,666 | 497 | 19 | 215 |
| Camden | 4,211 | 231 | 2,666 | 1,545 | 58 | 670 |
| Union | 9,345 | 231 | 2,666 | 6,679 | 251 | 2,894 |
| Edison | 6,186 | 231 | 2,666 | 3,520 | 132 | 1,525 |
| Trenton | 7,386 | 231 | 2,666 | 4,720 | 177 | 2,045 |
| *New Mexico* | | | | | | |
| Albuquerque | 4,234 | 12 | 2,447 | 1,787 | 73 | 15,491 |
| Las Cruces | 2,529 | 12 | 2,447 | 82 | 3 | 715 |
| Roswell | 2,620 | 12 | 2,447 | 173 | 7 | 1,503 |
| Santa Fe | 2,625 | 12 | 2,447 | 178 | 7 | 1,547 |
| *New York* | | | | | | |
| Albany | 3,500 | 63 | 2,498 | 1,002 | 40 | 1,578 |
| Buffalo | 3,866 | 63 | 2,498 | 1,368 | 55 | 2,155 |
| Elmira | 4,022 | 63 | 2,498 | 1,524 | 61 | 2,401 |
| Hempstead | 7,872 | 63 | 2,498 | 5,374 | 215 | 8,467 |
| Poughkeepsie | 3,433 | 63 | 2,498 | 935 | 37 | 1,473 |
| Rochester | 4,897 | 63 | 2,498 | 2,399 | 96 | 3,780 |
| Rome | 3,520 | 63 | 2,498 | 1,022 | 41 | 1,610 |
| Schenectady | 3,563 | 63 | 2,498 | 1,065 | 43 | 1,677 |
| Syracuse | 4,861 | 63 | 2,498 | 2,363 | 95 | 3,723 |
| *North Carolina* | | | | | | |
| Charlotte | 3,058 | 90 | 2,525 | 533 | 21 | 592 |
| Durham | 2,843 | 90 | 2,525 | 318 | 13 | 353 |
| Fayetteville | 2,136 | 90 | 2,525 | −389 | −15 | −432 |
| Greensboro | 3,837 | 90 | 2,525 | 1,312 | 52 | 1,458 |
| Raleigh | 3,677 | 90 | 2,525 | 1,152 | 46 | 1,280 |
| Rocky Mount | 3,257 | 90 | 2,525 | 732 | 29 | 813 |
| Winston-Salem | 3,558 | 90 | 2,525 | 1,033 | 41 | 1,148 |
| *North Dakota* | | | | | | |
| Fargo | 2,290 | 24 | 2,459 | −169 | −7 | −698 |
| *Ohio* | | | | | | |
| Akron | 5,005 | 108 | 2,543 | 2,462 | 97 | 2,270 |
| Canton | 3,000 | 108 | 2,543 | 457 | 18 | 421 |
| Cincinnati | 5,932 | 108 | 2,543 | 3,389 | 133 | 3,124 |
| Cleveland | 6,172 | 108 | 2,543 | 3,629 | 143 | 3,345 |
| Columbus | 4,733 | 108 | 2,543 | 2,190 | 86 | 2,019 |
| Dayton | 4,709 | 108 | 2,543 | 2,166 | 85 | 1,997 |
| Elyria | 4,056 | 108 | 2,543 | 1,513 | 59 | 1,395 |
| Lancaster | 2,750 | 108 | 2,543 | 207 | 8 | 190 |
| Lima | 3,481 | 108 | 2,543 | 938 | 37 | 864 |
| Mansfield | 3,145 | 108 | 2,543 | 602 | 24 | 555 |
| Sandusky | 4,340 | 108 | 2,543 | 1,797 | 71 | 1,656 |
| Springfield | 2,975 | 108 | 2,543 | 432 | 17 | 398 |
| Toledo | 5,083 | 108 | 2,543 | 2,540 | 100 | 2,341 |
| Youngstown | 3,480 | 108 | 2,543 | 937 | 37 | 863 |
| *Oklahoma* | | | | | | |
| Bartlesville | 3,360 | 42 | 2,477 | 883 | 36 | 2,106 |
| Enid | 2,588 | 42 | 2,477 | 111 | 4 | 265 |
| Lawton | 2,250 | 42 | 2,477 | −227 | −9 | −541 |
| Norman | 2,842 | 42 | 2,477 | 365 | 15 | 871 |
| Oklahoma City | 3,555 | 42 | 2,477 | 1,078 | 44 | 2,572 |
| Tulsa | 4,071 | 42 | 2,477 | 1,594 | 64 | 3,802 |

Appendix Table A–3. (*Continued*)

| State and city | Price per lot[a] (1) | Farm value per lot[b] (2) | Farm value plus improvement cost (2) + 2,435 (3) | Appreciation (1) − (3) (4) | Per cent appreciation over costs (4) ÷ (3) · 100 (5) | Per cent appreciation over farm value (4) ÷ (2) · 100 (6) |
|---|---|---|---|---|---|---|
| *Oregon* | | | | | | |
| Eugene | $3,223 | $ 38 | $2,473 | $ 750 | 30% | 1,969% |
| Portland | 3,612 | 38 | 2,473 | 1,139 | 46 | 2,991 |
| Salem | 3,213 | 38 | 2,473 | 740 | 30 | 1,943 |
| *Pennsylvania* | | | | | | |
| Allentown | 3,441 | 85 | 2,520 | 921 | 37 | 1,078 |
| Altoona | 2,000 | 85 | 2,520 | −520 | −21 | −609 |
| Erie | 2,763 | 85 | 2,520 | 243 | 10 | 284 |
| Harrisburg | 3,186 | 85 | 2,520 | 666 | 26 | 780 |
| Johnstown | 2,025 | 85 | 2,520 | −495 | −20 | −580 |
| Lancaster | 2,817 | 85 | 2,520 | 297 | 12 | 347 |
| Philadelphia | 4,469 | 85 | 2,520 | 1,949 | 77 | 2,282 |
| Pittsburgh | 5,508 | 85 | 2,520 | 2,988 | 119 | 3,499 |
| Reading | 1,775 | 85 | 2,520 | −745 | −30 | −873 |
| Williamsport | 2,800 | 85 | 2,520 | 280 | 11 | 327 |
| York | 2,524 | 85 | 2,520 | 4 | 0 | 4 |
| *Rhode Island* | | | | | | |
| Providence | 2,586 | 160 | 2,595 | −9 | 0 | −6 |
| *South Carolina* | | | | | | |
| Charleston | 3,425 | 63 | 2,498 | 927 | 37 | 1,479 |
| Columbia | 3,581 | 63 | 2,498 | 1,083 | 43 | 1,728 |
| Greenville | 3,186 | 63 | 2,498 | 688 | 28 | 1,098 |
| Spartanburg | 2,750 | 63 | 2,498 | 252 | 10 | 402 |
| *South Dakota* | | | | | | |
| Rapid City | 2,600 | 23 | 2,458 | 412 | 6 | 603 |
| Sioux Falls | 3,627 | 23 | 2,458 | 1,169 | 48 | 4,981 |
| *Tennessee* | | | | | | |
| Chattanooga | 4,066 | 63 | 2,498 | 1,568 | 63 | 2,470 |
| Jackson | 3,200 | 63 | 2,498 | 702 | 28 | 1,105 |
| Johnson City | 4,250 | 63 | 2,498 | 1,752 | 70 | 2,760 |
| Knoxville | 2,613 | 63 | 2,498 | 115 | 5 | 180 |
| Memphis | 3,805 | 63 | 2,498 | 1,307 | 52 | 2,059 |
| Nashville | 2,583 | 63 | 2,498 | 85 | 3 | 133 |
| Oak Ridge | 6,833 | 63 | 2,498 | 4,335 | 173 | 6,830 |
| *Texas* | | | | | | |
| Abilene | 2,190 | 42 | 2,477 | −287 | −12 | −690 |
| Amarillo | 2,306 | 42 | 2,477 | −171 | −7 | −411 |
| Austin | 3,200 | 42 | 2,477 | 723 | 29 | 1,742 |
| Beaumont | 2,573 | 42 | 2,477 | 96 | 4 | 232 |
| Corpus Christi | 3,355 | 42 | 2,477 | 878 | 35 | 2,115 |
| Dallas | 4,487 | 42 | 2,477 | 2,010 | 81 | 4,840 |
| El Paso | 3,055 | 42 | 2,477 | 578 | 23 | 1,393 |
| Fort Worth | 3,471 | 42 | 2,477 | 994 | 40 | 2,394 |
| Houston | 4,472 | 42 | 2,477 | 1,995 | 81 | 4,804 |
| Longview | 2,217 | 42 | 2,477 | −260 | −10 | −625 |
| Lubbock | 2,413 | 42 | 2,477 | −64 | −3 | −153 |
| Midland | 3,000 | 42 | 2,477 | 523 | 21 | 1,260 |
| San Angelo | 2,133 | 42 | 2,477 | −344 | −14 | −827 |
| San Antonio | 3,668 | 42 | 2,477 | 1,191 | 48 | 2,868 |
| Texarkana | 1,300 | 42 | 2,477 | −1,177 | −48 | −2,832 |
| Tyler | 2,625 | 42 | 2,477 | 148 | 6 | 357 |
| Victoria | 2,398 | 42 | 2,477 | −79 | −3 | −189 |
| Waco | 1,936 | 42 | 2,477 | −541 | −22 | −1,301 |
| Wichita Falls | 2,683 | 42 | 2,477 | 206 | 8 | 497 |
| *Utah* | | | | | | |
| Ogden | 3,242 | 26 | 2,461 | 781 | 32 | 2,986 |
| Salt Lake City | 3,993 | 26 | 2,461 | 1,532 | 62 | 5,857 |

Appendix Table A-3.   (*Continued*)

| State and city | Price per lot[a] | Farm value per lot[b] | Farm value plus im- provement cost (2) + 2,435 | Appre- ciation (1) − (3) | Per cent appreciation over costs (4) ÷ (3) · 100 | Per cent appreciation over farm value (4) ÷ (2) · 100 |
|---|---|---|---|---|---|---|
| | (1) | (2) | (3) | (4) | (5) | (6) |
| *Vermont* | | | | | | |
| Burlington | $3,000 | $ 33 | $2,468 | $ 532 | 22% | 1,589% |
| *Virginia* | | | | | | |
| Charlottesville | 6,200 | 63 | 2,498 | 3,702 | 148 | 5,869 |
| Danville | 2,627 | 63 | 2,498 | 129 | 5 | 204 |
| Lynchburg | 2,333 | 63 | 2,498 | −165 | −7 | −262 |
| Newport News | 3,975 | 63 | 2,498 | 1,477 | 59 | 2,341 |
| Norfolk | 3,530 | 63 | 2,498 | 1,032 | 41 | 1,636 |
| Petersburg | 2,647 | 63 | 2,498 | 149 | 6 | 236 |
| Richmond | 4,377 | 63 | 2,498 | 1,879 | 75 | 2,979 |
| Roanoke | 2,413 | 63 | 2,498 | −85 | −3 | −135 |
| *Washington* | | | | | | |
| Seattle | 3,909 | 56 | 2,491 | 1,418 | 57 | 2,543 |
| Spokane | 2,775 | 56 | 2,491 | 284 | 11 | 510 |
| Tacoma | 2,698 | 56 | 2,491 | 207 | 8 | 372 |
| Yakima | 3,940 | 56 | 2,491 | 1,449 | 58 | 2,599 |
| *West Virginia* | | | | | | |
| Charleston | 5,375 | 33 | 2,468 | 2,907 | 118 | 8,788 |
| Huntington | 3,550 | 33 | 2,468 | 1,082 | 44 | 3,271 |
| Wheeling | 3,500 | 33 | 2,468 | 1,032 | 42 | 3,120 |
| *Wisconsin* | | | | | | |
| Appleton | 3,833 | 84 | 2,519 | 1,314 | 52 | 1,567 |
| Beloit | 2,022 | 84 | 2,519 | −497 | −20 | −593 |
| Green Bay | 2,200 | 84 | 2,519 | −319 | −13 | −380 |
| Madison | 8,153 | 84 | 2,519 | 5,634 | 224 | 6,720 |
| Milwaukee | 5,713 | 84 | 2,519 | 3,194 | 127 | 3,810 |
| Racine | 4,117 | 84 | 2,519 | 1,598 | 63 | 1,906 |
| *Wyoming* | | | | | | |
| Cheyenne | 2,600 | 10 | 2,445 | 155 | 6 | 1,550 |
| High | 15,469 | 231 | 2,666 | 12,857 | 492 | 16,345 |
| Low | 1,300 | 10 | 2,445 | −1,177 | −48 | −2,832 |
| Average | 3,874 | 84 | 2,519 | 1,356 | 53 | 1,875 |

[a] Lot price data from National Association of Home Builders, *Economic News Notes*, Special Report 65–8 (Washington).

[b] Farm value per acre as shown in Appendix Table A–7 divided by a constant 2.6 lots per acre. A slight improvement in this variable might be obtained by using the FHA average lot sizes shown in Table 6. However, since the data are not available for all the cities in this table and FHA districts may not match exactly with the NAHB cities, a constant size was used.

Appendix Table A–4.   Future Capital Values and Rents to Support Present Appreciation
Values in 1964 (discounting at 6 per cent)

| City and state | New lots appre- ciation, 1964 | Future value | | Yearly rent to support value shown in | | Value of existing lots 10 yrs. old in 1964 |
| | | In 5 yrs. | In 10 yrs. | Column (2) | Column (3) | |
| --- | --- | --- | --- | --- | --- | --- |
| | (1) | (2) | (3) | (4) | (5) | (6) |
| Atlanta, Ga. | $2,543 | $3,403 | $4,555 | $266 | $356 | $2,570 |
| Austin, Texas | 723 | 968 | 1,296 | 76 | 101 | 1,951 |
| Charlotte, N.C. | 533 | 713 | 955 | 56 | 75 | 2,813 |
| Cincinnati, Ohio | 3,389 | 4,534 | 6,069 | 355 | 475 | 3,042 |
| Columbus, Ohio | 2,190 | 2,930 | 3,921 | 229 | 307 | 2,336 |
| Cleveland, Ohio | 3,629 | 4,855 | 6,499 | 380 | 508 | 2,986 |
| Dallas, Texas | 2,010 | 2,690 | 3,601 | 210 | 282 | 2,184 |
| Dayton, Ohio | 2,166 | 2,897 | 3,878 | 227 | 303 | 2,483 |
| Des Moines, Iowa | 1,237 | 1,655 | 2,216 | 129 | 173 | 2,601 |
| Detroit, Mich. | 3,246 | 4,343 | 5,814 | 340 | 455 | 2,339 |
| Fort Wayne, Ind. | 1,104 | 1,478 | 1,978 | 116 | 155 | 2,128 |
| Greensboro, N.C. | 1,312 | 1,755 | 2,350 | 137 | 184 | 2,453 |
| Indianapolis, Ind. | 2,507 | 3,355 | 4,491 | 262 | 351 | 2,075 |
| Lansing, Mich. | 1,781 | 2,383 | 3,190 | 186 | 249 | 2,207 |
| Memphis, Tenn. | 1,307 | 1,748 | 2,340 | 137 | 183 | 2,535 |
| Minneapolis, Minn. | 1,796 | 2,404 | 3,217 | 188 | 252 | 3,208 |
| Milwaukee, Wisc. | 3,194 | 4,274 | 5,721 | 334 | 447 | 3,871 |
| Omaha, Neb. | 1,276 | 1,707 | 2,285 | 134 | 179 | 2,548 |
| Oklahoma City, Okla. | 1,078 | 1,442 | 1,931 | 113 | 151 | 2,032 |
| Orlando, Fla. | 836 | 1,118 | 1,497 | 87 | 117 | 2,730 |
| Portland, Oregon | 1,139 | 1,524 | 2,040 | 119 | 160 | 2,473 |
| San Diego, Cal. | 5,227 | 6,994 | 9,362 | 547 | 732 | 5,004 |
| St. Louis, Mo. | 3,888 | 5,202 | 6,963 | 407 | 544 | 2,974 |
| Syracuse, N.Y. | 2,363 | 3,161 | 4,231 | 247 | 331 | 2,475 |
| Springfield, Mass. | 554 | 741 | 992 | 58 | 78 | 2,050 |
| Tacoma, Wash. | 207 | 277 | 371 | 22 | 29 | 2,281 |
| Washington, D.C. | 4,460 | 5,967 | 7,987 | 467 | 625 | 3,669 |

*Notes:* Column (1) is the appreciation in marginal fringe site value existing in 1964 above the cost of improvements and agricultural opportunity costs. This data was computed in Appendix Table A–3 for various cities
based on NAHB lot prices. It represents a value unsupported by present travel savings (rent). Column (2) is
the capital value that must be realized in five years for the present value in column (1) to be supported. In other
words, column (1) is the present value of column (2) to be received as a lump sum in five years. Column (2) =
(1964 appreciation × 1.338). Column (3) is similar to (2) except computed to produce a value not to be realized
for ten years. Column (3) = (1964 appreciation × 1.791).

Column (4) is the yearly rent or travel saving over a 25-year time horizon which the property must earn (beginning in five years) in order to support the capital value shown in column (2). Column (4) = Future value
as shown in column (2) = .0782. Column (5) = Future value as shown in column (3) × .0782. All discounting
in the table is at 6 per cent.

Column (6) is the actual 1964 value of existing lots with improvements that are approximately ten years old
on the average. The data are from the FHA, Division of Research and Statistics, Table 8–M.

A more detailed computation could be made to reflect the fact that in actuality as new areas are developed
the current fringe sites begin to earn some rent, which increases each year. They do not have to wait in five-year
intervals as indicated here. This example should be sufficient, however, to illustrate the principle involved.

Appendix Table A-5.  Actual Prices of Ten-Year-Old Lots, 1960, and Expected
Values Based on 1950 Prices

(*Dollars*)

| City and state | Market price of new sites, 1950 | Estimated capital value, 1960 (1) × 1.791 | Actual price of 10-year-old sites, 1960 |
|---|---|---|---|
| | (1) | (2) | (3) |
| Akron, Ohio | 813 | 1,456 | 2,255 |
| Albany–Schenectady–Troy, N.Y. | 723 | 1,295 | 1,771 |
| Albuquerque, N. Mex. | 725 | 1,299 | 2,360 |
| Amarillo, Tex. | 825 | 1,478 | 1,561 |
| Atlanta, Ga. | 965 | 1,728 | 2,194 |
| Atlantic City, N.J. | 916 | 1,641 | 1,913 |
| Austin, Tex. | 1,031 | 1,847 | 1,961 |
| Baltimore, Md. | 996 | 1,784 | 1,807 |
| Baton Rouge, La. | 1,235 | 2,212 | 2,937 |
| Beaumont–Port Arthur, Tex. | 1,080 | 1,934 | 1,955 |
| Birmingham, Ala. | 1,117 | 2,001 | 2,244 |
| Boston, Mass. | 826 | 1,479 | 2,377 |
| Buffalo, N.Y. | 832 | 1,490 | 2,142 |
| Charleston, S.C. | 1,019 | 1,825 | 2,182 |
| Charleston, W. Va. | 1,143 | 2,047 | 2,702 |
| Charlotte, N.C. | 1,052 | 1,884 | 2,403 |
| Chattanooga, Tenn. | 891 | 1,596 | 1,559 |
| Chicago, Ill. | 1,116 | 1,999 | 3,237 |
| Cincinnati, Ohio | 1,434 | 2,568 | 2,753 |
| Cleveland, Ohio | 919 | 1,646 | 2,608 |
| Columbus, Ohio | 1,022 | 1,830 | 2,051 |
| Corpus Christi, Tex. | 1,135 | 2,033 | 1,745 |
| Dallas, Tex. | 988 | 1,770 | 1,843 |
| Davenport–Rock Island–Moline, Ill. | 801 | 1,435 | 2,389 |
| Dayton, Ohio | 967 | 1,732 | 2,002 |
| Denver, Colo. | 838 | 1,501 | 2,606 |
| Des Moines, Iowa | 925 | 1,657 | 2,023 |
| Detroit, Mich. | 822 | 1,472 | 2,150 |
| El Paso, Tex. | 654 | 1,171 | 1,757 |
| Flint, Mich. | 793 | 1,420 | 2,538 |
| Fort Wayne, Ind. | 1,018 | 1,823 | 1,872 |
| Fort Worth, Tex. | 747 | 1,338 | 1,565 |
| Fresno, Calif. | 846 | 1,515 | 2,238 |
| Galveston–Texas City, Tex. | 1,244 | 2,228 | 1,882 |
| Grand Rapids, Mich. | 809 | 1,449 | 1,981 |
| Greensboro–High Point, N.C. | 737 | 1,320 | 2,277 |
| Greenville, S.C. | 812 | 1,454 | 1,807 |
| Hamilton–Middletown, Ohio | 1,069 | 1,915 | 2,183 |
| Harrisburg, Pa. | 959 | 1,718 | 1,560 |
| Hartford, Conn. | 928 | 1,662 | 2,356 |
| Honolulu, Hawaii | 4,303 | 7,707 | 6,917 |
| Houston, Tex. | 1,449 | 2,595 | 2,256 |
| Indianapolis, Ind. | 843 | 1,510 | 1,819 |
| Jackson, Miss. | 681 | 1,220 | 1,822 |
| Jacksonville, Fla. | 896 | 1,605 | 1,864 |
| Kansas City, Mo. | 1,079 | 1,933 | 1,848 |
| Knoxville, Tenn. | 749 | 1,342 | 1,681 |
| Lansing, Mich. | 802 | 1,436 | 1,990 |
| Lexington, Ky. | 1,150 | 2,060 | 2,644 |
| Lincoln, Neb. | 868 | 1,555 | 2,225 |

Appendix Table A–5. (*Continued*)

| City and state | Market price of new sites, 1950 | Estimated capital value, 1960 (1) × 1.791 | Actual price of 10-year-old sites, 1960 |
|---|---|---|---|
| | (1) | (2) | (3) |
| Little Rock, Ark. | 1,202 | 2,153 | 2,408 |
| Lorain–Elyria, Ohio | 875 | 1,567 | 2,157 |
| Los Angeles, Calif. | 1,514 | 2,712 | 3,745 |
| Louisville, Ky. | 944 | 1,691 | 2,499 |
| Lubbock, Tex. | 901 | 1,614 | 1,657 |
| Macon, Ga. | 728 | 1,304 | 1,851 |
| Memphis, Tenn. | 889 | 1,592 | 2,212 |
| Miami, Fla. | 1,224 | 2,192 | 3,088 |
| Milwaukee, Wis. | 1,104 | 1,977 | 3,825 |
| Minneapolis, Minn. | 734 | 1,315 | 2,601 |
| Mobile, Ala. | 1,289 | 2,309 | 2,517 |
| Montgomery, Ala. | 1,243 | 2,226 | 2,625 |
| Nashville, Tenn. | 924 | 1,655 | 2,255 |
| Newark, N.J. | 1,154 | 2,067 | 3,462 |
| New Haven, Conn. | 817 | 1,463 | 2,576 |
| New Orleans, La. | 1,803 | 3,229 | 4,052 |
| New York, N.Y. | 1,124 | 2,013 | 3,543 |
| Norfolk–Portsmouth, Va. | 741 | 1,327 | 1,872 |
| Ogden, Utah | 818 | 1,465 | 2,061 |
| Oklahoma City, Okla. | 1,143 | 2,047 | 1,758 |
| Omaha, Neb. | 802 | 1,436 | 2,192 |
| Orlando, Fla. | 1,001 | 1,793 | 2,519 |
| Philadelphia, Pa. | 927 | 1,660 | 1,611 |
| Phoenix, Ariz. | 912 | 1,633 | 2,333 |
| Pittsburgh, Pa. | 1,271 | 2,276 | 2,491 |
| Portland, Me. | 650 | 1,164 | 1,204 |
| Portland, Ore. | 670 | 1,200 | 1,885 |
| Providence, R.I. | 827 | 1,481 | 1,385 |
| Raleigh, N.C. | 690 | 1,236 | 2,108 |
| Richmond, Va. | 739 | 1,324 | 1,926 |
| Roanoke, Va. | 848 | 1,519 | 1,967 |
| Rochester, N.Y. | 811 | 1,453 | 2,027 |
| Sacramento, Calif. | 1,069 | 1,915 | 3,450 |
| St. Louis, Mo. | 1,490 | 2,669 | 2,398 |
| Salt Lake City, Utah | 868 | 1,555 | 2,268 |
| San Antonio, Tex. | 1,025 | 1,836 | 1,633 |
| San Bernardino, Calif. | 883 | 1,582 | 2,543 |
| San Diego, Calif. | 1,344 | 2,407 | 4,438 |
| San Francisco, Calif. | 1,339 | 2,398 | 3,200 |
| San Jose, Calif. | 1,365 | 2,445 | 3,810 |
| Savannah, Ga. | 876 | 1,569 | 2,285 |
| Seattle, Wash. | 1,072 | 1,920 | 2,660 |
| Shreveport, La. | 1,140 | 2,042 | 2,128 |
| Sioux Falls, S.D. | 928 | 1,662 | 1,654 |
| South Bend, Ind. | 895 | 1,603 | 1,717 |
| Spokane, Wash. | 532 | 953 | 1,700 |
| Springfield, Ill. | 798 | 1,429 | 2,330 |
| Springfield–Chicopee–Holyoke, Mass. | 695 | 1,245 | 1,545 |
| Stockton, Calif. | 1,112 | 1,992 | 2,589 |
| Syracuse, N.Y. | 801 | 1,435 | 1,931 |

Appendix Table A–5.   (*Continued*)

| City and state | Market price of new sites, 1950 | Estimated capital value, 1960 (1) × 1.791 | Actual price of 10-year-old sites, 1960 |
|---|---|---|---|
| | (1) | (2) | (3) |
| Tacoma, Wash. | 756 | 1,354 | 1,510 |
| Tampa–St. Petersburg, Fla. | 849 | 1,521 | 2,597 |
| Toledo, Ohio | 888 | 1,590 | 2,241 |
| Trenton, N.J. | 685 | 1,227 | 1,545 |
| Tulsa, Okla. | 1,072 | 1,920 | 1,852 |
| Washington, D.C. | 1,474 | 2,640 | 2,868 |
| Wichita, Kan. | 710 | 1,272 | 1,713 |
| Wichita Falls, Tex. | 823 | 1,474 | 1,278 |
| Wilmington, Del. | 1,081 | 1,936 | 2,092 |
| Winston-Salem, N.C. | 647 | 1,159 | 2,189 |
| Youngstown–Warren, Ohio | 810 | 1,451 | 2,078 |

*Note:* Price of site of 1950 new houses is extended at 6% for 10 years. Column (2) is computed in such a way that column (1) is the present value in 1950 of the future value of column (2) to be realized in 10 years (or 1960). This can be compared to the price of 10-year-old sites in 1960.

*Source: FHA Homes in Standard Metropolitan Areas, 1950*, and Federal Housing Administration, Division of Research and Statistics, Table 43–M, "Property Characteristics, Selected Housing Areas, One Family Homes," Sec. 203, 1960.

Appendix Table A–6.   Ratio of Existing Lot Prices to New Lot Prices, 1955 and 1963

| City and state | 1955 | 1963 | Negative (−) or positive (+) change bet. 1955 & 1963 |
|---|---|---|---|
| Akron, Ohio | .975 | n.a. | n.a. |
| Albany–Schenectady–Troy, N.Y. | 1.266 | .941 | − |
| Albuquerque, N. Mex. | 1.188 | .977 | − |
| Amarillo, Tex. | n.a. | .810 | n.a. |
| Atlanta, Ga. | 1.213 | .960 | − |
| Atlantic City, N.J. | n.a. | 1.213 | n.a. |
| Augusta, Ga. | n.a. | .843 | n.a. |
| Austin, Tex. | n.a. | 1.016 | n.a. |
| Bakersfield, Calif. | n.a. | .845 | n.a. |
| Baltimore, Md. | 1.083 | .861 | − |
| Baton Rouge, La. | n.a. | 1.086 | n.a. |
| Beaumont–Port Arthur, Tex. | n.a. | .958 | n.a. |
| Billings, Mont. | n.a. | 1.161 | n.a. |
| Birmingham, Ala. | 1.135 | 1.069 | − |
| Buffalo, N.Y. | .964 | .884 | − |
| Charleston, S.C. | n.a. | .940 | n.a. |
| Charlotte, N.C. | 1.196 | 1.093 | − |
| Chattanooga, Tenn. | n.a. | .802 | n.a. |
| Chicago, Ill. | .947 | 1.025 | + |
| Cincinnati, Ohio | n.a. | .853 | n.a. |
| Cleveland, Ohio | .870 | n.a. | n.a. |
| Columbia, S.C. | n.a. | 1.084 | n.a. |
| Columbus, Ohio | .822 | .743 | − |
| Corpus Christi, Tex. | n.a. | .874 | n.a. |
| Dallas, Tex. | 1.093 | 1.039 | − |
| Dayton, Ohio | .852 | .772 | − |
| Denver, Colo. | 1.127 | 1.027 | − |
| Des Moines, Iowa | 1.343 | n.a. | n.a. |
| Detroit, Mich. | .917 | .799 | − |
| El Paso, Tex. | 1.002 | .826 | − |
| Flint, Mich. | .952 | .855 | − |
| Ft. Lauderdale–Hollywood, Fla. | n.a. | 1.045 | n.a. |
| Fort Wayne, Ind. | .844 | .786 | − |
| Fort Worth, Tex. | .994 | .959 | − |
| Fresno, Calif. | 1.161 | n.a. | n.a. |
| Gary–Hammond–E. Chicago, Ind. | n.a. | .781 | n.a. |
| Grand Rapids, Mich. | 1.079 | .889 | − |
| Greensboro–High Point, N.C. | n.a. | .974 | n.a. |
| Hartford, Conn. | 1.054 | 1.152 | + |
| Honolulu, Hawaii | n.a. | 1.087 | n.a. |
| Houston, Tex. | 1.145 | 1.066 | − |
| Indianapolis, Ind. | .981 | .706 | − |
| Jackson, Miss. | n.a. | .981 | n.a. |
| Jacksonville, Fla. | 1.126 | .874 | − |
| Kansas City, Mo. | .866 | .839 | − |
| Knoxville, Tenn. | 1.208 | .886 | − |
| Lansing, Mich. | n.a. | .676 | n.a. |
| Las Vegas, Nev. | n.a. | 1.033 | n.a. |
| Lexington, Ky. | n.a. | .966 | n.a. |
| Lincoln, Neb. | n.a. | .887 | n.a. |
| Little Rock–North Little Rock, Ark. | 1.174 | .988 | − |
| Lorain–Elyria, Ohio | n.a. | .956 | n.a. |
| Los Angeles–Long Beach, Calif. | 1.440 | .974 | − |
| Louisville, Ky. | .863 | .967 | + |

Appendix Table A–6.   (*Continued*)

| City and state | 1955 | 1963 | Negative (−) or positive (+) change bet. 1955 & 1963 |
|---|---|---|---|
| Lubbock, Tex. | n.a. | .896 | n.a. |
| Memphis, Tenn. | 1.077 | .989 | − |
| Miami, Fla. | 1.242 | .928 | − |
| Milwaukee, Wis. | 1.357 | .914 | − |
| Minneapolis–St. Paul, Minn. | 1.240 | 1.184 | − |
| Mobile, Ala. | n.a. | .951 | n.a. |
| Montgomery, Ala. | n.a. | .850 | n.a. |
| Nashville, Tenn. | n.a. | .977 | n.a. |
| New Orleans, La. | 1.303 | 1.077 | − |
| Newport News–Hampton, Va. | n.a. | .966 | n.a. |
| New York, N.Y. | 1.160 | 1.667 | + |
| Norfolk–Portsmouth, Va. | .960 | .825 | − |
| Oklahoma City, Okla. | .960 | .914 | − |
| Omaha, Neb. | 1.172 | .839 | − |
| Orlando, Fla. | n.a. | .971 | n.a. |
| Pensacola, Fla. | n.a. | .852 | n.a. |
| Philadelphia, Pa. | .752 | .751 | − |
| Phoenix, Ariz. | 1.041 | 1.072 | + |
| Pittsburgh, Pa. | .946 | .817 | − |
| Portland, Ore., Wash. | .996 | .948 | − |
| Providence–Pawtucket, R.I., Mass. | n.a. | .946 | n.a. |
| Reno, Nev. | n.a. | 1.077 | n.a. |
| Richmond, Va. | 1.093 | .888 | − |
| Roanoke, Va. | n.a. | .912 | n.a. |
| Rochester, N.Y. | .838 | n.a. | n.a. |
| Sacramento, Calif. | 1.119 | 1.025 | − |
| St. Louis, Mo., Ill. | .905 | .864 | − |
| Salt Lake City, Utah | 1.186 | 1.067 | − |
| San Antonio, Tex. | 1.091 | .881 | − |
| San Bernardino–Riverside–Ontario, Calif. | 1.180 | .798 | − |
| San Diego, Calif. | 1.226 | .956 | − |
| San Francisco–Oakland, Calif. | 1.083 | .943 | − |
| San Jose, Calif. | 1.227 | .938 | − |
| San Juan, Puerto Rico | n.a. | 1.253 | n.a. |
| Santa Barbara, Calif. | n.a. | .828 | n.a. |
| Seattle, Wash. | 1.277 | 1.203 | − |
| Shreveport, La. | 1.089 | 1.044 | − |
| South Bend, Ind. | .833 | .644 | − |
| Spokane, Wash. | 1.137 | .956 | − |
| Springfield–Chicopee–Holyoke, Mass. | n.a. | .919 | n.a. |
| Stockton, Calif. | 1.087 | .954 | − |
| Syracuse, N.Y. | 1.072 | .798 | − |
| Tacoma, Wash. | 1.211 | 1.008 | − |
| Tampa–St. Petersburg, Fla. | 1.314 | 1.064 | − |
| Toledo, Ohio | .944 | .711 | − |
| Topeka, Kan. | .922 | n.a. | n.a. |
| Tucson, Ariz. | n.a. | .934 | n.a. |
| Tulsa, Okla. | 1.070 | .798 | − |
| Washington, D.C. | 1.313 | .960 | − |
| Wichita, Kan. | .975 | .875 | − |
| Wilmington, Del., N.J. | n.a. | .702 | n.a. |
| Youngstown–Warren, Ohio | 1.143 | .739 | − |

n.a. = Not available.
*Source:* Federal Housing Administration, Division of Research and Statistics, Table 43–M.

Appendix Table A–7.   Suburban Raw Land Price Appreciation above Farm Land
Prices, U.S. Cities, 1960

| State and city | Suburban price/acre | Farm price/acre | Appreciation (1) − (2) | Per cent appreciation over farm value (3) ÷ (2) |
|---|---|---|---|---|
| | (1) | (2) | (3) | (4) |
| *Alabama* | | | | |
| Birmingham | $1,398 | $ 91 | $1,307 | 1,436% |
| Decatur | 1,000 | 91 | 909 | 999 |
| Dothan | 1,600 | 91 | 1,509 | 1,658 |
| Huntsville | 1,844 | 91 | 1,753 | 1,926 |
| Mobile | 1,867 | 91 | 1,776 | 1,952 |
| Montgomery | 1,957 | 91 | 1,866 | 2,051 |
| Tuscaloosa | 1,167 | 91 | 1,076 | 1,182 |
| *Arizona* | | | | |
| Phoenix | 3,911 | 48 | 3,863 | 8,048 |
| Tucson | 2,767 | 48 | 2,719 | 5,665 |
| *Arkansas* | | | | |
| Fort Smith | 1,283 | 113 | 1,170 | 1,035 |
| Little Rock | 1,817 | 113 | 1,704 | 1,508 |
| Pine Bluff | 2,667 | 113 | 2,554 | 2,260 |
| *California* | | | | |
| Berkeley | 5,523 | 360 | 5,163 | 1,434 |
| Fresno[a] | 3,000 | 360 | 2,640 | 733 |
| Los Angeles | 8,851 | 360 | 8,491 | 2,359 |
| Modesto | 2,817 | 360 | 2,457 | 683 |
| Sacramento[a] | 3,315 | 360 | 2,955 | 821 |
| San Bernardino[a] | 4,852 | 360 | 4,492 | 1,248 |
| San Diego[a] | 4,518 | 360 | 4,158 | 1,155 |
| San Francisco | 8,167 | 360 | 7,807 | 2,169 |
| San Mateo | 11,750 | 360 | 11,390 | 3,164 |
| Santa Barbara | 5,253 | 360 | 4,893 | 1,359 |
| Santa Clara | 9,837 | 360 | 9,477 | 2,633 |
| *Colorado* | | | | |
| Colorado Springs | 2,122 | 54 | 2,068 | 3,830 |
| Denver | 2,901 | 54 | 2,847 | 5,272 |
| Pueblo | 1,067 | 54 | 1,013 | 1,876 |
| *Connecticut* | | | | |
| Hamden | 1,593 | 446 | 1,147 | 257 |
| New London | 920 | 446 | 474 | 106 |
| West Hartford | 2,161 | 446 | 1,715 | 385 |
| *Delaware* | | | | |
| Wilmington[a] | 2,909 | 243 | 2,666 | 1,097 |
| *Washington, D.C.*[a] | 3,379 | 290 | 3,089 | 1,065 |
| *Florida* | | | | |
| Daytona Beach | 2,980 | 217 | 2,763 | 1,273 |
| Gainesville | 1,605 | 217 | 1,388 | 640 |
| Jacksonville[a] | 1,172 | 217 | 955 | 440 |
| Lakeland | 2,250 | 217 | 2,033 | 937 |
| Miami | 5,070 | 217 | 4,853 | 2,236 |
| Orlando[a] | 2,393 | 217 | 2,176 | 1,003 |
| Pensacola | 1,464 | 217 | 1,247 | 575 |
| St. Petersburg[a] | 4,157 | 217 | 3,940 | 1,816 |
| Sarasota | 1,600 | 217 | 1,383 | 637 |
| Tallahassee | 1,500 | 217 | 1,283 | 591 |
| Tampa[a] | 2,464 | 217 | 2,247 | 1,035 |

Appendix Table A–7.	(*Continued*)

| State and city | Suburban price/acre | Farm price/acre | Appreciation (1) − (2) | Per cent appreciation over farm value (3) ÷ (2) |
|---|---|---|---|---|
| | (1) | (2) | (3) | (4) |
| *Georgia* | | | | |
| Albany | $  870 | $ 99 | $  771 | 779% |
| Athens | 642 | 99 | 543 | 548 |
| Atlanta[a] | 1,753 | 99 | 1,654 | 1,671 |
| Augusta | 975 | 99 | 876 | 885 |
| Columbus | 1,475 | 99 | 1,376 | 1,390 |
| Macon | 904 | 99 | 805 | 813 |
| Rome | 558 | 99 | 459 | 464 |
| Savannah | 1,333 | 99 | 1,234 | 1,246 |
| *Idaho* | | | | |
| Boise | 500 | 112 | 388 | 346 |
| *Illinois* | | | | |
| Alton | 767 | 316 | 451 | 143 |
| Belleville | 1,250 | 316 | 934 | 296 |
| Bloomington | 1,500 | 316 | 1,184 | 375 |
| Champaign | 2,333 | 316 | 2,017 | 638 |
| Chicago | 4,758 | 316 | 4,442 | 1,406 |
| Danville | 2,000 | 316 | 1,684 | 533 |
| Decatur | 2,192 | 316 | 1,876 | 594 |
| Galesburg | 1,000 | 316 | 684 | 216 |
| Kankakee | 1,400 | 316 | 1,084 | 343 |
| Rock Island | 2,540 | 316 | 2,224 | 704 |
| Peoria | 2,333 | 316 | 2,017 | 638 |
| Rockford[a] | 2,000 | 316 | 1,684 | 533 |
| Springfield | 2,475 | 316 | 2,159 | 683 |
| *Indiana* | | | | |
| Anderson | 1,500 | 264 | 1,236 | 468 |
| Bloomington | 870 | 264 | 606 | 230 |
| Evansville | 2,300 | 264 | 2,036 | 771 |
| Fort Wayne | 1,104 | 264 | 840 | 318 |
| Hammond | 1,714 | 264 | 1,450 | 549 |
| Indianapolis | 1,583 | 264 | 1,319 | 500 |
| Lafayette | 1,550 | 264 | 1,286 | 487 |
| Marion | 500 | 264 | 236 | 89 |
| Muncie | 1,192 | 264 | 928 | 352 |
| New Albany | 900 | 264 | 636 | 241 |
| Richmond | 1,342 | 264 | 1,078 | 408 |
| South Bend | 989 | 264 | 725 | 275 |
| *Iowa* | | | | |
| Ames | 1,762 | 257 | 1,505 | 586 |
| Cedar Rapids | 2,009 | 257 | 1,752 | 682 |
| Council Bluffs | 1,750 | 257 | 1,493 | 581 |
| Davenport | 1,800 | 257 | 1,543 | 600 |
| Des Moines | 1,321 | 257 | 1,064 | 414 |
| Fort Dodge | 750 | 257 | 493 | 192 |
| Iowa City | 1,167 | 257 | 910 | 354 |
| Sioux City | 450 | 257 | 193 | 75 |
| *Kansas* | | | | |
| Hutchinson | 1,057 | 101 | 956 | 947 |
| Salina | 1,414 | 101 | 1,313 | 1,300 |
| Topeka | 1,000 | 101 | 899 | 890 |
| Wichita[a] | 986 | 101 | 885 | 876 |

Appendix Table A–7.　(*Continued*)

| State and city | Suburban price/acre | Farm price/acre | Appreciation (1) − (2) | Per cent appreciation over farm value (3) ÷ (2) |
|---|---|---|---|---|
| | (1) | (2) | (3) | (4) |
| *Kentucky* | | | | |
| Covington | $1,420 | $137 | $1,283 | 936% |
| Lexington | 2,388 | 137 | 2,251 | 1,643 |
| Louisville | 2,721 | 137 | 2,584 | 1,886 |
| Owensboro | 2,569 | 137 | 2,432 | 1,775 |
| Paducah | 2,850 | 137 | 2,713 | 1,980 |
| *Louisiana* | | | | |
| Alexandria | 2,033 | 173 | 1,860 | 1,075 |
| Baton Rouge | 2,875 | 173 | 2,702 | 1,562 |
| Lafayette | 2,125 | 173 | 1,952 | 1,128 |
| Lake Charles | 2,500 | 173 | 2,327 | 1,345 |
| New Orleans | 5,663 | 173 | 5,490 | 3,173 |
| Shreveport | 1,844 | 173 | 1,671 | 966 |
| *Maine* | | | | |
| Portland | 400 | 80 | 320 | 400 |
| *Maryland* | | | | |
| Baltimore | 3,011 | 290 | 2,721 | 938 |
| Cumberland | 1,300 | 290 | 1,010 | 348 |
| *Massachusetts* | | | | |
| Attleboro | 833 | 314 | 519 | 165 |
| Boston | 1,728 | 314 | 1,414 | 450 |
| Fall River | 1,500 | 314 | 1,186 | 378 |
| Lowell | 671 | 314 | 357 | 114 |
| Lawrence[a] | 1,000 | 314 | 686 | 218 |
| Springfield | 703 | 314 | 389 | 124 |
| Worcester | 829 | 314 | 515 | 164 |
| *Michigan* | | | | |
| Ann Arbor | 3,125 | 194 | 2,931 | 1,511 |
| Battle Creek | 1,258 | 194 | 1,064 | 548 |
| Detroit | 3,845 | 194 | 3,651 | 1,882 |
| Flint[a] | 650 | 194 | 456 | 235 |
| Grand Rapids | 1,580 | 194 | 1,386 | 714 |
| Jackson | 733 | 194 | 539 | 278 |
| Kalamazoo[a] | 833 | 194 | 639 | 329 |
| Lansing | 1,400 | 194 | 1,206 | 622 |
| Midland | 2,500 | 194 | 2,306 | 1,189 |
| Muskegon | 667 | 194 | 473 | 244 |
| Port Huron | 550 | 194 | 356 | 184 |
| Saginaw | 1,756 | 194 | 1,562 | 805 |
| *Minnesota* | | | | |
| Duluth | 2,010 | 155 | 1,855 | 1,197 |
| Minneapolis[a] | 1,515 | 155 | 1,360 | 877 |
| Rochester | 5,000 | 155 | 4,845 | 3,126 |
| St. Paul[a] | 1,545 | 155 | 1,390 | 897 |
| *Mississippi* | | | | |
| Greenville | 1,265 | 107 | 1,158 | 1,082 |
| Hattiesburg | 1,125 | 107 | 1,018 | 951 |
| Jackson[a] | 1,863 | 107 | 1,756 | 1,641 |

Appendix Table A-7.  (*Continued*)

| State and city | Suburban price/acre | Farm price/acre | Appreciation (1) − (2) | Per cent appreciation over farm value (3) ÷ (2) |
|---|---|---|---|---|
|  | (1) | (2) | (3) | (4) |
| *Missouri* | | | | |
| Columbia | $  800 | $115 | $  685 | 596% |
| Kansas City | 1,991 | 115 | 1,876 | 1,631 |
| Springfield | 1,742 | 115 | 1,627 | 1,415 |
| St. Joseph | 970 | 115 | 855 | 743 |
| St. Louis | 4,222 | 115 | 4,107 | 3,571 |
| *Montana* | | | | |
| Billings | 2,900 | 35 | 2,865 | 8,186 |
| Great Falls | 1,300 | 35 | 1,265 | 3,614 |
| *Nebraska* | | | | |
| Lincoln[a] | 1,179 | 90 | 1,089 | 1,210 |
| Omaha | 1,928 | 90 | 1,838 | 2,042 |
| *Nevada* | | | | |
| Las Vegas | 3,200 | 31 | 3,169 | 10,223 |
| Reno | 3,389 | 31 | 3,358 | 10,832 |
| *New Hampshire* | | | | |
| Nashua | 1,500 | 104 | 1,396 | 1,342 |
| Manchester | 1,750 | 104 | 1,646 | 1,583 |
| *New Jersey* | | | | |
| Atlantic City | 914 | 528 | 386 | 73 |
| Camden | 3,003 | 528 | 2,475 | 469 |
| Union | 6,719 | 528 | 6,191 | 1,173 |
| Edison | 3,685 | 528 | 3,157 | 598 |
| Trenton | 2,220 | 528 | 1,692 | 320 |
| *New Mexico* | | | | |
| Albuquerque | 3,478 | 24 | 3,454 | 14,392 |
| Las Cruces | 1,600 | 24 | 1,576 | 6,567 |
| Roswell | 2,217 | 24 | 2,193 | 9,138 |
| Santa Fe | 1,200 | 24 | 1,176 | 4,900 |
| *New York* | | | | |
| Albany | 779 | 145 | 634 | 437 |
| Buffalo | 1,353 | 145 | 1,208 | 833 |
| Elmira | 1,400 | 145 | 1,255 | 866 |
| Hempstead | 11,084 | 145 | 10,939 | 7,544 |
| Poughkeepsie | 1,029 | 145 | 884 | 610 |
| Rochester | 1,671 | 145 | 1,526 | 1,052 |
| Rome | 1,150 | 145 | 1,005 | 693 |
| Schenectady | 863 | 145 | 718 | 495 |
| Syracuse | 1,788 | 145 | 1,643 | 1,133 |
| *North Carolina* | | | | |
| Charlotte[a] | 1,861 | 186 | 1,675 | 901 |
| Durham | 2,475 | 186 | 2,289 | 1,231 |
| Fayetteville | 1,000 | 186 | 814 | 438 |
| Greensboro[a] | 1,409 | 186 | 1,223 | 658 |
| Raleigh[a] | 1,914 | 186 | 1,728 | 929 |
| Rocky Mount | 2,000 | 186 | 1,814 | 975 |
| Winston-Salem | 1,286 | 186 | 1,100 | 591 |
| *North Dakota* | | | | |
| Fargo | 950 | 53 | 897 | 1,692 |

Appendix Table A–7. (*Continued*)

| State and city | Suburban price/acre | Farm price/acre | Appreciation (1) − (2) | Per cent appreciation over farm value (3) ÷ (2) |
|---|---|---|---|---|
| | (1) | (2) | (3) | (4) |
| *Ohio* | | | | |
| Akron | $2,154 | $248 | $1,906 | 769% |
| Canton | 1,404 | 248 | 1,156 | 466 |
| Cincinnati | 2,479 | 248 | 2,231 | 900 |
| Cleveland | 2,083 | 248 | 1,835 | 740 |
| Columbus[a] | 2,561 | 248 | 2,313 | 933 |
| Dayton[a] | 2,185 | 248 | 1,937 | 781 |
| Elyria | 567 | 248 | 319 | 129 |
| Lancaster | 1,000 | 248 | 752 | 303 |
| Lima | 1,250 | 248 | 1,002 | 404 |
| Mansfield | 945 | 248 | 697 | 281 |
| Sandusky | 1,188 | 248 | 940 | 379 |
| Springfield | 1,400 | 248 | 1,152 | 465 |
| Toledo | 2,095 | 248 | 1,847 | 745 |
| Youngstown | 820 | 248 | 572 | 231 |
| *Oklahoma* | | | | |
| Bartlesville | 1,375 | 86 | 1,289 | 1,499 |
| Enid | 1,500 | 86 | 1,414 | 1,644 |
| Lawton | 1,338 | 86 | 1,252 | 1,456 |
| Norman | 1,613 | 86 | 1,527 | 1,776 |
| Oklahoma City | 1,948 | 86 | 1,862 | 2,165 |
| Tulsa | 2,572 | 86 | 2,486 | 2,891 |
| *Oregon* | | | | |
| Eugene | 1,992 | 88 | 1,904 | 2,164 |
| Portland | 4,718 | 88 | 4,630 | 5,261 |
| Salem | 1,140 | 88 | 1,052 | 1,195 |
| *Pennsylvania* | | | | |
| Allentown | 1,645 | 188 | 1,457 | 775 |
| Altoona | 1,000 | 188 | 812 | 432 |
| Erie | 1,150 | 188 | 962 | 512 |
| Harrisburg | 1,685 | 188 | 1,497 | 796 |
| Johnstown | 1,267 | 188 | 1,079 | 574 |
| Lancaster | 2,758 | 188 | 2,570 | 1,367 |
| Philadelphia | 1,850 | 188 | 1,662 | 884 |
| Pittsburgh | 3,078 | 188 | 2,890 | 1,537 |
| Reading | 1,043 | 188 | 855 | 455 |
| Williamsport | 3,025 | 188 | 2,837 | 1,509 |
| York | 1,093 | 188 | 905 | 481 |
| *Rhode Island* | | | | |
| Providence | 1,686 | 379 | 1,307 | 345 |
| *South Carolina* | | | | |
| Charleston | 1,900 | 137 | 1,763 | 1,287 |
| Columbia | 1,500 | 137 | 1,363 | 995 |
| Greenville | 1,050 | 137 | 913 | 666 |
| Spartanburg | 1,000 | 137 | 863 | 630 |
| *South Dakota* | | | | |
| Rapid City | 1,100 | 51 | 1,049 | 2,057 |
| Sioux Falls | 1,420 | 51 | 1,369 | 2,684 |

Appendix Table A–7.    (*Continued*)

| State and city | Suburban price/acre | Farm price/acre | Appreciation (1) − (2) | Per cent appreciation over farm value (3) ÷ (2) |
|---|---|---|---|---|
| | (1) | (2) | (3) | (4) |
| *Tennessee* | | | | |
| Chattanooga | $1,173 | $132 | $1,041 | 789% |
| Jackson | 1,000 | 132 | 868 | 658 |
| Johnson City | 1,000 | 132 | 868 | 658 |
| Knoxville | 1,039 | 132 | 907 | 687 |
| Memphis | 2,548 | 132 | 2,416 | 1,830 |
| Nashville | 1,280 | 132 | 1,148 | 870 |
| Oak Ridge | 300 | 132 | 168 | 127 |
| *Texas* | | | | |
| Abilene | 1,550 | 85 | 1,465 | 1,724 |
| Amarillo | 2,136 | 85 | 2,051 | 2,413 |
| Austin[a] | 1,509 | 85 | 1,424 | 1,675 |
| Beaumont | 2,436 | 85 | 2,351 | 2,766 |
| Corpus Christi | 1,917 | 85 | 1,832 | 2,155 |
| Dallas | 3,802 | 85 | 3,717 | 4,373 |
| El Paso | 2,079 | 85 | 1,994 | 2,346 |
| Fort Worth[a] | 1,535 | 85 | 1,450 | 1,706 |
| Houston | 2,709 | 85 | 2,624 | 3,087 |
| Longview | 933 | 85 | 848 | 998 |
| Lubbock | 3,000 | 85 | 2,915 | 3,429 |
| Midland | 3,000 | 85 | 2,915 | 3,429 |
| San Angelo | 1,075 | 85 | 990 | 1,165 |
| San Antonio | 2,099 | 85 | 2,014 | 2,369 |
| Texarkana | 900 | 85 | 815 | 959 |
| Tyler | 1,557 | 85 | 1,472 | 1,732 |
| Victoria | 1,580 | 85 | 1,495 | 1,759 |
| Waco | 1,120 | 85 | 1,035 | 1,218 |
| Wichita Falls | 2,067 | 85 | 1,982 | 2,332 |
| *Utah* | | | | |
| Ogden | 2,233 | 60 | 2,173 | 3,622 |
| Salt Lake City | 2,688 | 60 | 2,628 | 4,380 |
| *Vermont* | | | | |
| Burlington | 400 | 81 | 319 | 394 |
| *Virginia* | | | | |
| Charlottesville | 1,075 | 140 | 935 | 668 |
| Danville | 1,000 | 140 | 860 | 614 |
| Lynchburg | 563 | 140 | 423 | 302 |
| Newport News | 2,032 | 140 | 1,892 | 1,351 |
| Norfolk | 1,996 | 140 | 1,856 | 1,326 |
| Petersburg | 1,013 | 140 | 873 | 624 |
| Richmond | 1,768 | 140 | 1,628 | 1,163 |
| Roanoke | 1,113 | 140 | 973 | 695 |
| *Washington* | | | | |
| Seattle[a] | 2,394 | 133 | 2,261 | 1,700 |
| Spokane | 2,660 | 133 | 2,527 | 1,900 |
| Tacoma | 1,470 | 133 | 1,337 | 1,005 |
| Yakima | 2,500 | 133 | 2,367 | 1,780 |
| *West Virginia* | | | | |
| Charleston | 1,000 | 75 | 925 | 1,233 |
| Huntington | 2,100 | 75 | 2,025 | 2,700 |
| Wheeling | 1,300 | 75 | 1,225 | 1,633 |

Appendix Table A–7.  (*Continued*)

| State and city | Suburban price/acre | Farm price/acre | Appreciation (1) − (2) | Per cent appreciation over farm value (3) ÷ (2) |
|---|---|---|---|---|
| | (1) | (2) | (3) | (4) |
| *Wisconsin* | | | | |
| Appleton | $2,560 | $133 | $2,427 | 1,825% |
| Beloit | 696 | 133 | 563 | 423 |
| Green Bay | 1,700 | 133 | 1,567 | 1,178 |
| Madison[a] | 1,750 | 133 | 1,617 | 1,216 |
| Milwaukee[a] | 2,296 | 133 | 2,163 | 1,626 |
| Racine | 1,925 | 133 | 1,792 | 1,347 |
| *Wyomimg* | | | | |
| Cheyenne | 2,000 | 22 | 1,978 | 8,991 |
| High | 11,750 | 528 | 11,390 | 14,392 |
| Low | 400 | 22 | 168 | 75 |
| Average | 1,995 | 183 | 1,812 | 1,466 |
| Standard deviation | | | 1,507 | 1,821 |

[a] Cities which had higher than average change in population of the urban area and less than average per cent appreciation in land value. Data are not available for all cities in table.

*Source:* Suburban raw land prices from National Association of Home Builders, *Economic News Notes*, Special Report 65–8 (Washington). Farm prices are state averages from U.S. Department of Agriculture, *Farm Real Estate Market Developments*, CD–66, pp. 13–14.

Appendix Table A–8.   Suburban Raw Land Price Appreciation above Farm Land
Prices, U.S. Cities, 1964

| State and city | Suburban price/acre | Farm price/acre | Appreciation (1) − (2) | Per cent appreciation over farm value (3) ÷ (2) |
|---|---|---|---|---|
|  | (1) | (2) | (3) | (4) |
| *Alabama* |  |  |  |  |
| Birmingham | $1,994 | $116 | $1,878 | 1,619% |
| Decatur | 800 | 116 | 684 | 590 |
| Dothan | 1,800 | 116 | 1,684 | 1,452 |
| Huntsville | 2,378 | 116 | 2,262 | 1,950 |
| Mobile | 2,800 | 116 | 2,684 | 2,314 |
| Montgomery | 2,425 | 116 | 2,309 | 1,991 |
| Tuscaloosa | 2,333 | 116 | 2,217 | 1,911 |
| *Arizona* |  |  |  |  |
| Phoenix | 8,630 | 60 | 8,570 | 14,283 |
| Tucson | 4,000 | 60 | 3,940 | 6,567 |
| *Arkansas* |  |  |  |  |
| Fort Smith | 1,050 | 152 | 898 | 591 |
| Little Rock | 3,058 | 152 | 2,906 | 1,912 |
| Pine Bluff | 2,500 | 152 | 2,348 | 1,545 |
| *California* |  |  |  |  |
| Berkeley | 10,614 | 460 | 10,154 | 2,207 |
| Fresno | 4,500 | 460 | 4,040 | 878 |
| Los Angeles | 17,177 | 460 | 16,717 | 3,634 |
| Modesto | 3,725 | 460 | 3,265 | 710 |
| Sacramento | 5,646 | 460 | 5,186 | 1,127 |
| San Bernardino | 4,532 | 460 | 4,072 | 885 |
| San Diego | 7,730 | 460 | 7,270 | 1,580 |
| San Francisco | 13,536 | 460 | 13,076 | 2,843 |
| San Mateo | 21,589 | 460 | 21,129 | 4,593 |
| Santa Barbara | 10,913 | 460 | 10,453 | 2,272 |
| Santa Clara | 14,466 | 460 | 14,006 | 3,045 |
| *Colorado* |  |  |  |  |
| Colorado Springs | 3,192 | 64 | 3,128 | 4,888 |
| Denver | 3,872 | 64 | 3,808 | 5,950 |
| Pueblo | 1,525 | 64 | 1,461 | 2,283 |
| *Connecticut* |  |  |  |  |
| Hamden | 2,756 | 500 | 2,256 | 451 |
| New London | 1,425 | 500 | 925 | 185 |
| West Hartford | 4,505 | 500 | 4,005 | 801 |
| *Delaware* |  |  |  |  |
| Wilmington | 5,150 | 295 | 4,855 | 1,646 |
| *Washington, D.C.* | 5,785 | 365 | 5,420 | 1,485 |
| *Florida* |  |  |  |  |
| Daytona Beach | 5,363 | 307 | 5,056 | 1,647 |
| Gainesville | 2,044 | 307 | 1,737 | 566 |
| Jacksonville | 1,925 | 307 | 1,618 | 527 |
| Lakeland | 3,000 | 307 | 2,693 | 877 |
| Miami | 5,806 | 307 | 5,499 | 1,791 |
| Orlando | 3,089 | 307 | 2,782 | 906 |
| Pensacola | 2,261 | 307 | 1,954 | 636 |
| St. Petersburg | 4,518 | 307 | 4,211 | 1,372 |
| Sarasota | 2,800 | 307 | 2,493 | 812 |
| Tallahassee | 2,025 | 307 | 1,718 | 560 |
| Tampa | 3,562 | 307 | 3,255 | 1,060 |

Appendix Table A–8.    (*Continued*)

| State and city | Suburban price/acre | Farm price/acre | Appreciation (1) − (2) | Per cent appreciation over farm value (3) ÷ (2) |
|---|---|---|---|---|
| | (1) | (2) | (3) | (4) |
| *Georgia* | | | | |
| Albany | $1,260 | $127 | $1,133 | 892% |
| Athens | 1,000 | 127 | 873 | 687 |
| Atlanta | 2,476 | 127 | 2,349 | 1,850 |
| Augusta | 1,791 | 127 | 1,664 | 1,310 |
| Columbus | 2,438 | 127 | 2,311 | 1,820 |
| Macon | 1,163 | 127 | 1,036 | 816 |
| Rome | 933 | 127 | 806 | 635 |
| Savannah | 1,550 | 127 | 1,423 | 1,120 |
| *Idaho* | | | | |
| Boise | 750 | 125 | 625 | 500 |
| *Illinois* | | | | |
| Alton | 700 | 348 | 352 | 101 |
| Belleville | 1,683 | 348 | 1,335 | 384 |
| Bloomington | 2,000 | 348 | 1,652 | 475 |
| Champaign | 3,933 | 348 | 3,585 | 1,030 |
| Chicago | 6,961 | 348 | 6,613 | 1,900 |
| Danville | 3,440 | 348 | 3,092 | 889 |
| Decatur | 2,300 | 348 | 1,952 | 561 |
| Galesburg | 2,200 | 348 | 1,852 | 532 |
| Kankakee | 1,200 | 348 | 852 | 245 |
| Rock Island | 3,836 | 348 | 3,488 | 1,002 |
| Peoria | 3,400 | 348 | 3,052 | 877 |
| Rockford | 2,492 | 348 | 2,144 | 616 |
| Springfield | 2,200 | 348 | 1,852 | 532 |
| *Indiana* | | | | |
| Anderson | 2,157 | 293 | 1,864 | 636 |
| Bloomington | 1,640 | 293 | 1,347 | 460 |
| Evansville | 2,833 | 293 | 2,540 | 867 |
| Fort Wayne | 1,615 | 293 | 1,322 | 451 |
| Hammond | 2,000 | 293 | 1,707 | 583 |
| Indianapolis | 2,852 | 293 | 2,559 | 873 |
| Lafayette | 2,125 | 293 | 1,832 | 625 |
| Marion | 1,000 | 293 | 707 | 241 |
| Muncie | 1,042 | 293 | 749 | 256 |
| New Albany | 1,550 | 293 | 1,257 | 429 |
| Richmond | 2,000 | 293 | 1,707 | 583 |
| South Bend | 1,202 | 293 | 909 | 310 |
| *Iowa* | | | | |
| Ames | 2,392 | 265 | 2,127 | 803 |
| Cedar Rapids | 2,710 | 265 | 2,445 | 923 |
| Council Bluffs | 1,625 | 265 | 1,360 | 513 |
| Davenport | 2,175 | 265 | 1,910 | 721 |
| Des Moines | 1,785 | 265 | 1,520 | 574 |
| Fort Dodge | 4,000 | 265 | 3,735 | 1,409 |
| Iowa City | 2,580 | 265 | 2,315 | 874 |
| *Kansas* | | | | |
| Hutchinson | 2,156 | 114 | 2,042 | 1,791 |
| Salina | 1,875 | 114 | 1,761 | 1,545 |
| Topeka | 1,550 | 114 | 1,436 | 1,260 |
| Wichita | 1,678 | 114 | 1,564 | 1,372 |

Appendix Table A–8.    (*Continued*)

| State and city | Suburban price/acre | Farm price/acre | Appreciation (1) − (2) | Per cent appreciation over farm value (3) ÷ (2) |
|---|---|---|---|---|
| | (1) | (2) | (3) | (4) |
| *Kentucky* | | | | |
| Covington | $2,988 | $171 | $2,817 | 1,647% |
| Lexington | 3,300 | 171 | 3,129 | 1,830 |
| Louisville | 3,784 | 171 | 3,613 | 2,113 |
| Owensboro | 2,765 | 171 | 2,594 | 1,517 |
| Paducah | 11,500 | 171 | 11,329 | 6,625 |
| *Louisiana* | | | | |
| Alexandria | 3,275 | 213 | 3,062 | 1,438 |
| Baton Rouge | 3,667 | 213 | 3,454 | 1,622 |
| Lafayette | 2,720 | 213 | 2,507 | 1,177 |
| Lake Charles | 3,000 | 213 | 2,787 | 1,308 |
| New Orleans | 9,746 | 213 | 9,533 | 4,476 |
| Shreveport | 2,607 | 213 | 2,394 | 1,124 |
| *Maine* | | | | |
| Portland | 567 | 93 | 474 | 510 |
| *Maryland* | | | | |
| Baltimore | 4,939 | 365 | 4,574 | 1,253 |
| Cumberland | 1,100 | 365 | 735 | 201 |
| *Massachusetts* | | | | |
| Attleboro | 1,000 | 349 | 651 | 187 |
| Boston | 2,868 | 349 | 2,519 | 722 |
| Fall River | 2,300 | 349 | 1,951 | 559 |
| Lowell | 1,025 | 349 | 676 | 194 |
| Lawrence | 1,650 | 349 | 1,301 | 373 |
| Springfield | 1,731 | 349 | 1,382 | 396 |
| Worcester | 1,186 | 349 | 837 | 240 |
| *Michigan* | | | | |
| Ann Arbor | 4,375 | 218 | 4,157 | 1,907 |
| Battle Creek | 1,740 | 218 | 1,522 | 698 |
| Detroit | 6,861 | 218 | 6,643 | 3,047 |
| Flint | 921 | 218 | 703 | 322 |
| Grand Rapids | 2,220 | 218 | 2,002 | 918 |
| Jackson | 1,050 | 218 | 832 | 382 |
| Kalamazoo | 1,200 | 218 | 982 | 450 |
| Lansing | 2,563 | 218 | 2,345 | 1,076 |
| Midland | 2,750 | 218 | 2,532 | 1,161 |
| Muskegon | 750 | 218 | 532 | 244 |
| Port Huron | 617 | 218 | 399 | 183 |
| Saginaw | 2,675 | 218 | 2,457 | 1,127 |
| *Minnesota* | | | | |
| Duluth | 850 | 168 | 682 | 406 |
| Minneapolis | 2,160 | 168 | 1,992 | 1,186 |
| Rochester | 5,000 | 168 | 4,832 | 2,876 |
| St. Paul | 2,675 | 168 | 2,507 | 1,492 |
| *Mississippi* | | | | |
| Greenville | 1,960 | 135 | 1,825 | 1,352 |
| Hattiesburg | 1,625 | 135 | 1,490 | 1,104 |
| Jackson | 2,665 | 135 | 2,530 | 1,874 |

Appendix Table A–8.  (*Continued*)

| State and city | Suburban price/acre (1) | Farm price/acre (2) | Appreciation (1) − (2) (3) | Per cent appreciation over farm value (3) ÷ (2) (4) |
|---|---|---|---|---|
| *Missouri* | | | | |
| Columbia | $1,500 | $139 | $1,361 | 979% |
| Kansas City | 3,103 | 139 | 2,964 | 2,132 |
| Springfield | 2,128 | 139 | 1,989 | 1,431 |
| St. Joseph | 1,413 | 139 | 1,274 | 917 |
| St. Louis | 4,673 | 139 | 4,534 | 3,262 |
| *Montana* | | | | |
| Billings | 2,200 | 42 | 2,158 | 5,138 |
| Great Falls | 1,500 | 42 | 1,458 | 3,471 |
| *Nebraska* | | | | |
| Lincoln | 1,846 | 104 | 1,742 | 1,675 |
| Omaha | 2,821 | 104 | 2,717 | 2,613 |
| *Nevada* | | | | |
| Las Vegas | 5,621 | 36 | 5,585 | 15,514 |
| Reno | 6,586 | 36 | 6,550 | 18,194 |
| *New Hampshire* | | | | |
| Nashua | 2,500 | 118 | 2,382 | 2,019 |
| Manchester | 2,500 | 118 | 2,382 | 2,019 |
| *New Jersey* | | | | |
| Atlantic City | 1,993 | 600 | 1,393 | 232 |
| Camden | 3,820 | 600 | 3,220 | 537 |
| Union | 12,061 | 600 | 11,461 | 1,910 |
| Edison | 7,015 | 600 | 6,415 | 1,069 |
| Trenton | 2,783 | 600 | 2,183 | 364 |
| *New Mexico* | | | | |
| Albuquerque | 4,558 | 30 | 4,528 | 15,093 |
| Las Cruces | 2,216 | 30 | 2,186 | 7,287 |
| Roswell | 2,750 | 30 | 2,720 | 9,067 |
| Santa Fe | 2,000 | 30 | 1,970 | 6,567 |
| *New York* | | | | |
| Albany | 4,630 | 165 | 4,465 | 2,706 |
| Buffalo | 2,687 | 165 | 2,522 | 1,528 |
| Elmira | 3,071 | 165 | 2,906 | 1,761 |
| Hempstead | 15,161 | 165 | 14,996 | 9,088 |
| Poughkeepsie | 2,114 | 165 | 1,949 | 1,181 |
| Rochester | 2,249 | 165 | 2,084 | 1,263 |
| Rome | 1,350 | 165 | 1,185 | 718 |
| Schenectady | 829 | 165 | 664 | 402 |
| Syracuse | 2,600 | 165 | 2,435 | 1,476 |
| *North Carolina* | | | | |
| Charlotte | 2,633 | 234 | 2,399 | 1,025 |
| Durham | 2,533 | 234 | 2,299 | 982 |
| Fayetteville | 1,506 | 234 | 1,272 | 544 |
| Greensboro | 1,970 | 234 | 1,736 | 742 |
| Raleigh | 2,438 | 234 | 2,204 | 942 |
| Rocky Mount | 1,850 | 234 | 1,616 | 691 |
| Winston-Salem | 3,019 | 234 | 2,785 | 1,190 |
| *North Dakota* | | | | |
| Fargo | 2,250 | 63 | 2,187 | 3,471 |

Appendix Table A–8.    (*Continued*)

| State and city | Suburban price/acre | Farm price/acre | Appreciation (1) − (2) | Per cent appreciation over farm value (3) ÷ (2) |
|---|---|---|---|---|
| | (1) | (2) | (3) | (4) |
| *Ohio* | | | | |
| Akron | $2,132 | $282 | $1,850 | 656% |
| Canton | 1,913 | 282 | 1,631 | 578 |
| Cincinnati | 3,696 | 282 | 3,414 | 1,211 |
| Cleveland | 3,022 | 282 | 2,740 | 972 |
| Columbus | 3,710 | 282 | 3,428 | 1,216 |
| Dayton | 2,780 | 282 | 2,498 | 886 |
| Elyria | 1,567 | 282 | 1,285 | 456 |
| Lancaster | 1,150 | 282 | 868 | 308 |
| Lima | 2,150 | 282 | 1,868 | 662 |
| Mansfield | 1,056 | 282 | 774 | 274 |
| Sandusky | 2,125 | 282 | 1,843 | 654 |
| Springfield | 1,583 | 282 | 1,301 | 461 |
| Toledo | 3,431 | 282 | 3,149 | 1,117 |
| Youngstown | 1,840 | 282 | 1,558 | 552 |
| *Oklahoma* | | | | |
| Bartlesville | 2,075 | 109 | 1,966 | 1,804 |
| Enid | 1,933 | 109 | 1,824 | 1,673 |
| Lawton | 2,000 | 109 | 1,891 | 1,735 |
| Norman | 1,756 | 109 | 1,647 | 1,511 |
| Oklahoma City | 3,228 | 109 | 3,119 | 2,861 |
| Tulsa | 3,928 | 109 | 3,819 | 3,504 |
| *Oregon* | | | | |
| Eugene | 3,101 | 99 | 3,002 | 3,032 |
| Portland | 4,078 | 99 | 3,979 | 4,019 |
| Salem | 1,767 | 99 | 1,668 | 1,685 |
| *Pennsylvania* | | | | |
| Allentown | 3,755 | 222 | 3,533 | 1,591 |
| Altoona | 1,500 | 222 | 1,278 | 576 |
| Erie | 1,650 | 222 | 1,428 | 643 |
| Harrisburg | 2,255 | 222 | 2,033 | 916 |
| Johnstown | 1,750 | 222 | 1,528 | 688 |
| Lancaster | 4,233 | 222 | 4,011 | 1,807 |
| Philadelphia | 4,992 | 222 | 4,770 | 2,149 |
| Pittsburgh | 4,114 | 222 | 3,892 | 1,753 |
| Reading | 2,125 | 222 | 1,903 | 857 |
| Williamsport | 3,800 | 222 | 3,578 | 1,612 |
| York | 1,319 | 222 | 1,097 | 494 |
| *Rhode Island* | | | | |
| Providence | 1,295 | 417 | 878 | 211 |
| *South Carolina* | | | | |
| Charleston | 3,094 | 163 | 2,931 | 1,798 |
| Columbia | 3,586 | 163 | 3,423 | 2,100 |
| Greenville | 1,456 | 163 | 1,293 | 793 |
| Spartanburg | 1,750 | 163 | 1,587 | 974 |
| *South Dakota* | | | | |
| Rapid City | 1,250 | 61 | 1,189 | 1,949 |
| Sioux Falls | 2,114 | 61 | 2,053 | 3,366 |

Appendix Table A–8. (*Continued*)

| State and city | Suburban price/acre | Farm price/acre | Appreciation (1) − (2) | Per cent appreciation over farm value (3) ÷ (2) |
|---|---|---|---|---|
| | (1) | (2) | (3) | (4) |
| *Tennessee* | | | | |
| Chattanooga | $1,669 | $165 | $1,504 | 912% |
| Jackson | 2,500 | 165 | 2,335 | 1,415 |
| Johnson City | 1,733 | 165 | 1,568 | 950 |
| Knoxville | 1,359 | 165 | 1,194 | 724 |
| Memphis | 3,681 | 165 | 3,516 | 2,131 |
| Nashville | 1,588 | 165 | 1,423 | 862 |
| Oak Ridge | 6,650 | 165 | 6,485 | 3,930 |
| *Texas* | | | | |
| Abilene | 1,000 | 108 | 892 | 826 |
| Amarillo | 2,446 | 108 | 2,338 | 2,165 |
| Austin | 2,373 | 108 | 2,265 | 2,097 |
| Beaumont | 2,958 | 108 | 2,850 | 2,639 |
| Corpus Christi | 2,543 | 108 | 2,435 | 2,255 |
| Dallas | 7,277 | 108 | 7,169 | 6,638 |
| El Paso | 2,638 | 108 | 2,530 | 2,343 |
| Fort Worth | 2,549 | 108 | 2,441 | 2,260 |
| Houston | 4,095 | 108 | 3,987 | 3,692 |
| Longview | 1,333 | 108 | 1,225 | 1,134 |
| Lubbock | 4,000 | 108 | 3,892 | 3,604 |
| Midland | 3,000 | 108 | 2,892 | 2,678 |
| San Angelo | 1,438 | 108 | 1,330 | 1,231 |
| San Antonio | 2,279 | 108 | 2,171 | 2,010 |
| Texarkana | 1,325 | 108 | 1,217 | 1,127 |
| Tyler | 1,940 | 108 | 1,832 | 1,696 |
| Victoria | 1,864 | 108 | 1,756 | 1,626 |
| Waco | 1,780 | 108 | 1,672 | 1,548 |
| Wichita Falls | 1,500 | 108 | 1,392 | 1,289 |
| *Utah* | | | | |
| Ogden | 3,256 | 68 | 3,188 | 4,688 |
| Salt Lake City | 4,018 | 68 | 3,950 | 5,809 |
| *Vermont* | | | | |
| Burlington | 400 | 87 | 313 | 360 |
| *Virginia* | | | | |
| Charlottesville | 3,242 | 164 | 3,078 | 1,877 |
| Danville | 1,600 | 164 | 1,436 | 876 |
| Lynchburg | 588 | 164 | 424 | 259 |
| Newport News | 3,229 | 164 | 3,065 | 1,869 |
| Norfolk | 4,100 | 164 | 3,936 | 2,400 |
| Petersburg | 2,040 | 164 | 1,876 | 1,144 |
| Richmond | 2,720 | 164 | 2,556 | 1,559 |
| Roanoke | 1,500 | 164 | 1,336 | 815 |
| *Washington* | | | | |
| Seattle | 3,881 | 145 | 3,736 | 2,577 |
| Spokane | 2,071 | 145 | 1,926 | 1,328 |
| Tacoma | 2,770 | 145 | 2,625 | 1,810 |
| Yakima | 3,750 | 145 | 3,605 | 2,486 |
| *West Virginia* | | | | |
| Charleston | 1,200 | 86 | 1,114 | 1,295 |
| Huntington | 2,340 | 86 | 2,254 | 2,621 |
| Wheeling | 1,600 | 86 | 1,514 | 1,760 |

Appendix Table A–8.    (*Continued*)

| State and city | Suburban price/acre | Farm price/acre | Appreciation (1) − (2) | Per cent appreciation over farm value (3) ÷ (2) |
|---|---|---|---|---|
|  | (1) | (2) | (3) | (4) |
| *Wisconsin* |  |  |  |  |
| Appleton | $2,600 | $218 | $2,382 | 1,093% |
| Beloit | 1,250 | 218 | 1,032 | 473 |
| Green Bay | 2,250 | 218 | 2,032 | 932 |
| Madison | 6,218 | 218 | 6,000 | 2,752 |
| Milwaukee | 3,791 | 218 | 3,573 | 1,639 |
| Racine | 2,813 | 218 | 2,595 | 1,190 |
| *Wyoming* |  |  |  |  |
| Cheyenne | 3,000 | 26 | 2,974 | 11,438 |
| High | 21,589 | 600 | 21,129 | 18,194 |
| Low | 400 | 26 | 313 | 101 |
| Average | 3,030 | 217 | 2,812 | 1,819 |

*Source:* Suburban raw land prices from National Association of Home Builders, *Economic News Notes*, Special Report 65–8 (Washington). Farm prices are state averages from U.S. Department of Agriculture, *Farm Real Estate Market Developments*, CD–66, pp. 13–14.

## Appendix Table A–9. Housing Area Characteristics, U.S. Cities, 1960

| State and city | Land area (sq. mi.) | Urbanized area | | | | | | SMSA | City | | |
| | | Pop. | Δ Pop. 1950–60 (%) | Δ Land area, 1950–60 (%) | Pop./ sq. mi. | Income ($) | Fringe pop. (%) | Pop. in ring, work in city (%) | Pop. | Δ Pop., 1950–60 (%) | Income ($) |
| | (7)a | (8) | (3) | (4) | (9) | (5) | (6) | (13) | (12) | (11) | (10) |
| *Alabama* | | | | | | | | | | | |
| Birmingham | 156.8 | 521,330 | 17.1 | 56 | 3,325 | 5,143 | 35 | 42 | 340,887 | 4.6 | 4,947 |
| Decatur | n.a. | n.a. | n.a. | n.a. | n.a. | n.a. | n.a. | n.a. | 29,217 | 46.3 | 5,261 |
| Dothan | n.a. | n.a. | n.a. | n.a. | n.a. | n.a. | n.a. | n.a. | 31,440 | 45.7 | 4,347 |
| Huntsville | 53.2 | 74,970 | n.a. | n.a. | 1,409 | 6,368 | 3 | n.a. | 72,365 | 340.3 | 6,313 |
| Mobile | 171.5 | 268,139 | 46.6 | 318 | 1,563 | 5,239 | 24 | 42 | 202,779 | 57.2 | 5,452 |
| Montgomery | 39.2 | 142,893 | 30.5 | 37 | 3,645 | 5,089 | 6 | 50 | 134,393 | 26.2 | 5,065 |
| Tuscaloosa | 30.5 | 76,815 | n.a. | n.a. | 2,519 | 4,678 | 18 | n.a. | 63,370 | 36.6 | 4,620 |
| *Arizona* | | | | | | | | | | | |
| Phoenix | 248.4 | 552,043 | 155.5 | 351 | 2,222 | 6,138 | 20 | 26 | 439,170 | 311.1 | 6,117 |
| Tucson | 86.4 | 227,433 | n.a. | n.a. | 2,632 | 5,654 | 6 | 38 | 212,892 | 368.4 | 5,703 |
| *Arkansas* | | | | | | | | | | | |
| Fort Smith | 29.3 | 61,640 | 10.0 | 4 | 2,104 | 4,482 | 14 | n.a. | 52,991 | 10.5 | 4,644 |
| Little Rock | 62.2 | 185,017 | 20.4 | 62 | 2,975 | 5,092 | 42 | 51 | 107,813 | 5.5 | 5,234 |
| Pine Bluff | n.a. | n.a. | n.a. | n.a. | n.a. | n.a. | n.a. | n.a. | 44,037 | 18.5 | 4,604 |
| *California* | | | | | | | | | | | |
| Berkeley | n.a. | n.a. | n.a. | n.a. | n.a. | n.a. | n.a. | n.a. | 111,268 | −2.2 | 6,576 |
| Fresno | 60.6 | 213,444 | 63.4 | 99 | 3,522 | 6,112 | 37 | 31 | 133,929 | 46.1 | 6,109 |
| Los Angeles | 1,370.0 | 6,488,791 | 62.3 | 57 | 4,736 | 7,073 | 62 | 30 | 2,479,015 | 25.8 | 6,896 |
| Modesto | n.a. | n.a. | n.a. | n.a. | n.a. | n.a. | n.a. | n.a. | 36,585 | 110.4 | 6,357 |
| Sacramento | 134.0 | 451,920 | 113.4 | 222 | 3,373 | 7,088 | 58 | 35 | 191,667 | 39.3 | 6,943 |
| San Bernardino | 169.4 | 377,531 | 178.1 | 180 | 2,229 | 6,185 | 76 | 26 | 91,922 | 45.8 | 6,125 |
| San Diego | 275.7 | 836,175 | 93.1 | 108 | 3,033 | 6,706 | 31 | 26 | 573,224 | 71.4 | 6,614 |
| San Francisco | 568.5 | 2,430,663 | 20.2 | n.a. | 4,276 | 7,073 | 70 | 26 | 740,316 | −4.5 | 6,717 |
| San Mateo | n.a. | n.a. | n.a. | n.a. | n.a. | n.a. | n.a. | n.a. | 69,870 | 67.2 | 8,236 |
| Santa Barbara | 29.7 | 72,740 | n.a. | n.a. | 2,449 | 6,610 | 19 | n.a. | 58,768 | 31.0 | 6,477 |
| Santa Clara | n.a. | n.a. | n.a. | n.a. | n.a. | n.a. | n.a. | n.a. | 58,880 | 403.2 | 7,472 |
| *Colorado* | | | | | | | | | | | |
| Colorado Springs | 29.3 | 100,220 | n.a. | n.a. | 3,420 | 5,624 | 30 | 60 | 70,194 | 54.4 | 5,669 |
| Denver | 166.6 | 803,624 | 61.1 | 58 | 4,824 | 6,601 | 39 | 40 | 493,887 | 18.8 | 6,361 |
| Pueblo | 25.5 | 103,336 | 41.1 | 61 | 4,052 | 5,552 | 12 | 42 | 91,181 | 43.2 | 5,698 |

| | | | | | | | | | | | |
|---|---|---|---|---|---|---|---|---|---|---|---|
| *Connecticut* | | | | | | | | | | | |
| Hamden | n.a. | n.a. | n.a. | n.a. | n.a. | n.a. | n.a. | n.a. | 41,056 | 38.2 | n.a. |
| New London | n.a. | n.a. | n.a. | n.a. | n.a. | n.a. | n.a. | n.a. | 34,182 | 11.9 | 6,098 |
| West Hartford | 131.2 | 381,619 | 26.9 | 148 | 2,909 | 7,211 | 84 | 33 | 62,382 | 40.5 | n.a. |
| *Delaware* | | | | | | | | | | | |
| Wilmington | 81.4 | 283,667 | 51.4 | 74 | 3,485 | 6,894 | 66 | 28 | 95,827 | −13.2 | 5,589 |
| Washington, D.C. | 340.7 | 1,808,423 | 40.7 | 91 | 5,308 | 7,603 | 59 | 44 | 736,956 | −4.8 | 5,993 |
| *Florida* | | | | | | | | | | | |
| Daytona Beach | n.a. | n.a. | n.a. | n.a. | n.a. | n.a. | n.a. | n.a. | 37,395 | 23.9 | 3,986 |
| Gainesville | n.a. | n.a. | n.a. | n.a. | n.a. | n.a. | n.a. | n.a. | 29,701 | 10.6 | 4,702 |
| Jacksonville | 111.4 | 372,569 | 53.4 | 119 | 3,344 | 5,365 | 46 | 52 | 201,030 | −1.7 | 4,433 |
| Lakeland | n.a. | n.a. | n.a. | n.a. | n.a. | n.a. | n.a. | n.a. | 41,350 | 34.0 | 4,902 |
| Miami | 183.1 | 852,705 | 85.9 | 57 | 4,657 | 5,401 | 66 | 38 | 291,688 | 17.0 | 4,450 |
| Orlando | 76.8 | 200,995 | 174.7 | 208 | 2,617 | 5,413 | 56 | 32 | 88,135 | 68.3 | 5,037 |
| Pensacola | 45.8 | 128,049 | n.a. | n.a. | 2,796 | 5,308 | 56 | 28 | 56,752 | 30.5 | 5,204 |
| St. Petersburg | 115.2 | 324,842 | 183.5 | 65 | 2,820 | 4,332 | 44 | 31 | 181,298 | 87.4 | 4,232 |
| Sarasota | n.a. | n.a. | n.a. | n.a. | n.a. | n.a. | n.a. | n.a. | 34,083 | 80.4 | 4,558 |
| Tallahassee | n.a. | n.a. | n.a. | n.a. | n.a. | n.a. | n.a. | n.a. | 48,174 | 76.9 | 4,889 |
| Tampa | 103.4 | 301,790 | 68.3 | 153 | 2,919 | 4,749 | 9 | 31 | 274,970 | 120.5 | 4,667 |
| *Georgia* | | | | | | | | | | | |
| Albany | 24.4 | 58,353 | n.a. | n.a. | 2,392 | 4,490 | 4 | n.a. | 55,890 | 79.4 | 4,467 |
| Athens | n.a. | n.a. | n.a. | n.a. | n.a. | n.a. | 37 | 46 | 31,355 | 11.3 | 4,025 |
| Atlanta | 245.8 | 768,125 | 51.2 | 133 | 3,125 | 5,844 | 43 | 22 | 487,455 | 47.1 | 5,029 |
| Augusta | 43.1 | 123,698 | 41.0 | 144 | 2,870 | 4,484 | 26 | 46 | 70,626 | −1.2 | 3,603 |
| Columbus | 53.8 | 158,382 | 33.7 | 53 | 2,944 | 4,332 | 39 | 30 | 116,779 | 46.7 | 4,267 |
| Macon | 33.2 | 114,161 | 22.1 | 51 | 3,439 | 4,852 | n.a. | n.a. | 69,764 | −0.7 | 4,228 |
| Rome | n.a. | n.a. | n.a. | n.a. | n.a. | n.a. | 12 | 35 | 32,226 | 8.8 | 4,558 |
| Savannah | 61.1 | 169,887 | 32.5 | 172 | 2,780 | 4,860 | n.a. | n.a. | 149,245 | 24.7 | 4,761 |
| *Idaho* | | | | | | | | | | | |
| Boise | n.a. | n.a. | n.a. | n.a. | n.a. | n.a. | n.a. | n.a. | 34,481 | 0.3 | 5,851 |
| *Illinois* | | | | | | | | | | | |
| Alton | n.a. | n.a. | n.a. | n.a. | n.a. | n.a. | n.a. | n.a. | 43,047 | 32.2 | 6,453 |
| Belleville | n.a. | n.a. | n.a. | n.a. | n.a. | n.a. | n.a. | n.a. | 37,264 | 13.9 | 6,440 |
| Bloomington | n.a. | n.a. | n.a. | n.a. | n.a. | n.a. | 36 | n.a. | 36,271 | 6.2 | 5,944 |
| Champaign | 12.4 | 78,014 | n.a. | n.a. | 6,291 | 6,357 | 40 | 20 | 49,583 | 25.3 | 6,531 |
| Chicago | 959.8 | 5,959,213 | 21.1 | 50 | 6,209 | 7,292 | n.a. | 33 | 3,550,404 | −1.9 | 6,738 |
| Danville | n.a. | n.a. | n.a. | n.a. | n.a. | n.a. | 13 | n.a. | 41,856 | 10.5 | 5,812 |
| Decatur | 27.6 | 89,516 | 21.4 | 80 | 3,243 | 6,142 | n.a. | 55 | 78,004 | 17.7 | 6,054 |
| Galesburg | n.a. | n.a. | n.a. | n.a. | n.a. | n.a. | n.a. | n.a. | 37,243 | 18.5 | 6,272 |
| Kankakee | n.a. | n.a. | n.a. | n.a. | n.a. | n.a. | n.a. | n.a. | 27,666 | 7.0 | 6,054 |
| Rock Island | 95.9 | 227,176 | 16.5 | 75 | 2,369 | 6,617 | 77 | 35 | 51,863 | 6.5 | 6,558 |

Appendix Table A-9. (Continued)

| State and city | Urbanized area Land area (sq. mi.) (7)a | Pop. (8) | ΔPop. 1950-60 (%) (3) | ΔLand area, 1950-60 (%) (4) | Pop./ sq. mi. (9) | Income ($) (5) | Fringe pop. (%) (6) | SMSA Pop. in ring, work in city (%) (13) | City Pop. (12) | ΔPop., 1950-60 (%) (11) | Income ($) (10) |
|---|---|---|---|---|---|---|---|---|---|---|---|
| *Illinois (Continued)* | | | | | | | | | | | |
| Peoria | 50.4 | 181,432 | 17.4 | 52 | 3,600 | 6,387 | 43 | 28 | 103,162 | −7.8 | 5,961 |
| Rockford | 43.2 | 171,681 | 40.5 | 66 | 3,974 | 6,828 | 26 | 59 | 126,706 | 36.4 | 6,865 |
| Springfield | 32.6 | 111,403 | 14.4 | 90 | 3,417 | 6,177 | 25 | 44 | 83,271 | 2.0 | 6,007 |
| *Indiana* | | | | | | | | | | | |
| Anderson | n.a. | n.a. | n.a. | n.a. | n.a. | n.a. | n.a. | n.a. | 49,061 | 4.8 | 5,995 |
| Bloomington | n.a. | n.a. | n.a. | n.a. | n.a. | n.a. | n.a. | n.a. | 31,357 | 11.3 | 5,448 |
| Evansville | 34.1 | 143,660 | 4.4 | 54 | 4,213 | 5,297 | 1 | 32 | 141,543 | 10.0 | 5,299 |
| Fort Wayne | 48.6 | 179,571 | 28.0 | 117 | 3,695 | 6,571 | 10 | 53 | 161,766 | 21.1 | 6,492 |
| Hammond | n.a. | n.a. | n.a. | n.a. | n.a. | n.a. | n.a. | n.a. | 111,698 | 27.5 | 6,902 |
| Indianapolis | 44.9 | 639,340 | 27.3 | 60 | 4,412 | 6,534 | 26 | 59 | 476,258 | 11.5 | 6,106 |
| Lafayette | n.a. | n.a. | n.a. | n.a. | n.a. | n.a. | n.a. | n.a. | 42,330 | 19.0 | 6,193 |
| Marion | n.a. | n.a. | n.a. | n.a. | n.a. | n.a. | n.a. | n.a. | 37,854 | 25.8 | 5,890 |
| Muncie | 17.6 | 77,504 | n.a. | n.a. | 4,404 | 5,733 | 11 | 48 | 68,603 | 17.3 | 5,667 |
| New Albany | n.a. | n.a. | n.a. | n.a. | n.a. | n.a. | n.a. | n.a. | 37,812 | 28.8 | 5,439 |
| Richmond | n.a. | n.a. | n.a. | n.a. | n.a. | n.a. | n.a. | n.a. | 44,149 | 11.7 | 5,583 |
| South Bend | 64.0 | 218,933 | n.a. | n.a. | 3,421 | 6,615 | 40 | 44 | 132,445 | 14.3 | 6,682 |
| *Iowa* | | | | | | | | | | | |
| Ames | n.a. | n.a. | n.a. | n.a. | n.a. | n.a. | n.a. | n.a. | 27,003 | 17.9 | 6,192 |
| Cedar Rapids | 40.4 | 105,118 | 34.4 | 40 | 2,602 | 6,654 | 12 | 49 | 92,035 | 27.3 | 6,687 |
| Council Bluffs | n.a. | n.a. | n.a. | n.a. | n.a. | n.a. | n.a. | n.a. | 55,641 | 22.5 | 5,967 |
| Davenport | 95.9 | 227,176 | 16.5 | 75 | 2,369 | 6,612 | 61 | 35 | 88,981 | 19.4 | 6,479 |
| Des Moines | 97.0 | 241,115 | 20.6 | 43 | 2,486 | 6,529 | 13 | 45 | 208,982 | 17.4 | 6,436 |
| Fort Dodge | n.a. | n.a. | n.a. | n.a. | n.a. | n.a. | n.a. | n.a. | 28,399 | 13.1 | 6,059 |
| Iowa City | n.a. | n.a. | n.a. | n.a. | n.a. | n.a. | n.a. | n.a. | 33,443 | 22.9 | 5,769 |
| Sioux City | 54.3 | 97,926 | 8.7 | 10 | 1,803 | 5,800 | 9 | 16 | 89,159 | 6.2 | 5,812 |
| *Kansas* | | | | | | | | | | | |
| Hutchinson | n.a. | n.a. | n.a. | n.a. | n.a. | n.a. | n.a. | n.a. | 37,574 | 11.9 | 5,469 |
| Salina | n.a. | n.a. | n.a. | n.a. | n.a. | n.a. | n.a. | n.a. | 43,202 | 65.0 | 5,475 |
| Topeka | 36.2 | 119,500 | 34.1 | 112 | 3,301 | 6,040 | 0 | 37 | 119,484 | 51.6 | 6,039 |
| Wichita | 79.7 | 292,138 | 50.6 | 113 | 3,665 | 6,178 | 13 | 36 | 254,698 | 51.4 | 6,121 |

| | | | | | | | | | | | |
|---|---|---|---|---|---|---|---|---|---|---|---|
| *Kentucky* | | | | | | | | | | | |
| Covington | n.a. | n.a. | n.a. | n.a. | n.a. | n.a. | n.a. | n.a. | 60,376 | −6.3 | 5,302 |
| Lexington | 27.2 | 111,940 | n.a. | n.a. | 4,115 | 5,427 | 44 | 55 | 62,810 | 13.1 | 3,995 |
| Louisville | 135.6 | 606,659 | 28.3 | 104 | 4,474 | 5,734 | 36 | 38 | 390,639 | 5.8 | 5,280 |
| Owensboro | n.a. | n.a. | n.a. | n.a. | n.a. | n.a. | n.a. | n.a. | 42,471 | 26.2 | 5,076 |
| Paducah | n.a. | n.a. | n.a. | n.a. | n.a. | n.a. | n.a. | n.a. | 34,479 | 5.0 | 4,750 |
| *Louisiana* | | | | | | | | | | | |
| Alexandria | n.a. | n.a. | n.a. | n.a. | n.a. | n.a. | n.a. | n.a. | 40,279 | 15.4 | 3,768 |
| Baton Rouge | 56.8 | 193,485 | 39.3 | 38 | 3,406 | 5,733 | 21 | 62 | 152,419 | 21.3 | 5,789 |
| Lafayette | n.a. | n.a. | n.a. | n.a. | n.a. | n.a. | n.a. | n.a. | 40,400 | 20.4 | 4,361 |
| Lake Charles | 24.8 | 89,115 | 28.1 | n.a. | 3,593 | 5,133 | 29 | 34 | 63,392 | 53.6 | 5,462 |
| New Orleans | 266.5 | 845,237 | 39.9 | 20 | 3,172 | 5,202 | 26 | 42 | 627,525 | 10.0 | 4,807 |
| Shreveport | 52.4 | 208,583 | n.a. | 72 | 3,981 | 5,237 | 21 | 36 | 164,372 | 29.2 | 5,205 |
| *Maine* | | | | | | | | | | | |
| Portland | 51.2 | 111,701 | −1.6 | n.a. | 2,182 | 5,607 | 35 | 44 | 72,566 | −6.5 | 5,363 |
| *Maryland* | | | | | | | | | | | |
| Baltimore | 220.3 | 1,418,948 | 22.1 | 45 | 6,441 | 6,196 | 34 | 36 | 939,024 | −1.1 | 5,659 |
| Cumberland | n.a. | n.a. | n.a. | n.a. | n.a. | n.a. | n.a. | n.a. | 33,415 | −11.3 | 5,098 |
| *Massachusetts* | | | | | | | | | | | |
| Attleboro | n.a. | n.a. | n.a. | n.a. | n.a. | n.a. | n.a. | n.a. | 27,118 | 13.9 | 6,171 |
| Boston | 515.8 | 2,413,236 | 8.0 | 50 | 4,679 | 6,622 | 71 | 27 | 697,197 | −13.0 | 5,747 |
| Fall River | 47.6 | 123,951 | 4.9 | 35 | 2,604 | 5,168 | 19 | 49 | 99,942 | −10.7 | 4,970 |
| Lowell | 30.0 | 118,547 | 11.1 | 81 | 3,952 | 5,872 | 22 | 28 | 92,107 | −5.3 | 5,679 |
| Lawrence | 70.5 | 116,125 | 47.9 | 338 | 1,647 | 5,963 | 39 | 34 | 70,933 | −11.9 | 5,508 |
| Springfield | 238.8 | 449,777 | 26.0 | 43 | 1,883 | 6,182 | 61 | 31 | 174,463 | 7.4 | 5,994 |
| Worcester | 61.3 | 225,466 | 2.8 | 41 | 3,678 | 5,960 | 17 | 41 | 186,587 | −8.3 | 5,804 |
| *Michigan* | | | | | | | | | | | |
| Ann Arbor | 27.9 | 115,282 | n.a. | n.a. | 4,132 | 7,086 | 42 | 31 | 67,340 | 39.6 | 7,550 |
| Battle Creek | n.a. | n.a. | n.a. | n.a. | n.a. | n.a. | n.a. | n.a. | 44,169 | −9.2 | 6,029 |
| Detroit | 731.9 | 3,537,709 | 28.6 | 85 | 4,834 | 6,838 | 53 | 33 | 1,670,144 | −9.7 | 6,069 |
| Flint | 75.2 | 277,786 | 40.6 | 68 | 3,694 | 6,275 | 29 | 57 | 196,940 | 20.7 | 6,340 |
| Grand Rapids | 91.2 | 294,230 | 29.7 | 95 | 3,226 | 6,411 | 40 | 40 | 177,313 | 0.5 | 6,068 |
| Jackson | 22.1 | 71,412 | n.a. | n.a. | 3,231 | 6,580 | 29 | 46 | 50,720 | −0.7 | 6,422 |
| Kalamazoo | 42.1 | 115,659 | 38.8 | 99 | 2,747 | 6,576 | 29 | 50 | 82,089 | 42.3 | 6,365 |
| Lansing | 47.2 | 169,235 | 26.3 | 53 | 3,585 | 6,588 | 36 | 39 | 107,807 | 17.0 | 6,477 |
| Midland | n.a. | n.a. | n.a. | n.a. | n.a. | n.a. | n.a. | n.a. | 27,779 | 94.5 | 7,690 |
| Muskegon | 24.1 | 95,350 | 11.9 | 11 | 3,956 | 6,064 | 51 | 65 | 46,485 | −4.0 | 5,942 |
| Port Huron | n.a. | n.a. | n.a. | n.a. | n.a. | n.a. | n.a. | n.a. | 36,084 | 1.0 | 5,824 |
| Saginaw | 31.1 | 129,215 | 22.0 | 30 | 4,155 | 6,147 | 24 | 50 | 98,265 | 5.8 | 5,921 |

Appendix Table A-9. (Continued)

| | Urbanized area | | | | | | | SMSA | City | | |
|---|---|---|---|---|---|---|---|---|---|---|---|
| State and city | Land area (sq. mi.) | Pop. | Δ Pop. 1950-60 (%) | Δ Land area, 1950-60 (%) | Pop./ sq. mi. | Income ($) | Fringe pop. (%) | Pop. in ring, work in city (%) | Pop. | Δ Pop., 1950-60 (%) | Income ($) |
| | (7)a | (8) | (3) | (4) | (9) | (5) | (6) | (13) | (12) | (11) | (10) |
| *Minnesota* | | | | | | | | | | | |
| Duluth | 104.5 | 144,763 | 1.2 | n.a. | 1,385 | 5,826 | 26 | 17 | 106,884 | 2.3 | 5,877 |
| Minneapolis | 657.3 | 1,377,143 | 39.5 | 185 | 2,095 | 6,890 | 65 | 50 | 482,872 | -7.4 | 6,401 |
| Rochester | n.a. | n.a. | n.a. | n.a. | n.a. | n.a. | n.a. | n.a. | 40,663 | 36.1 | 6,638 |
| St. Paul | 657.3 | 1,377,143 | 39.5 | 185 | 2,095 | 6,890 | 77 | 50 | 313,411 | 0.7 | 6,543 |
| *Mississippi* | | | | | | | | | | | |
| Greenville | n.a. | n.a. | n.a. | n.a. | n.a. | n.a. | n.a. | n.a. | 41,502 | 49.2 | 3,812 |
| Hattiesburg | n.a. | n.a. | n.a. | n.a. | n.a. | n.a. | n.a. | n.a. | 34,989 | 18.7 | 4,232 |
| Jackson | 49.7 | 147,480 | 47.1 | 77 | 2,967 | 5,221 | 2 | 38 | 144,422 | 47.0 | 5,216 |
| *Missouri* | | | | | | | | | | | |
| Columbia | n.a. | n.a. | n.a. | n.a. | n.a. | n.a. | n.a. | n.a. | 36,650 | 14.6 | 5,616 |
| Kansas City | 282.4 | 921,121 | 31.9 | 90 | 3,262 | 6,381 | 48 | 40 | 475,539 | 4.1 | 5,906 |
| Springfield | 35.6 | 97,224 | 28.7 | 100 | 2,731 | 4,945 | 1 | 56 | 95,865 | 43.7 | 4,955 |
| St. Joseph | 28.8 | 81,187 | -1.3 | 67 | 2,819 | 5,501 | 2 | n.a. | 79,673 | 1.4 | 5,522 |
| St. Louis | 323.2 | 1,667,693 | 19.0 | 42 | 5,160 | 6,301 | 55 | 34 | 750,026 | -12.5 | 5,355 |
| *Montana* | | | | | | | | | | | |
| Billings | 15.5 | 60,712 | n.a. | n.a. | 3,917 | 6,447 | 13 | n.a. | 52,851 | 66.0 | 6,638 |
| Great Falls | 12.9 | 57,629 | n.a. | n.a. | 4,467 | 6,249 | 4 | n.a. | 55,244 | 41.2 | 6,257 |
| *Nebraska* | | | | | | | | | | | |
| Lincoln | 35.0 | 136,220 | 36.9 | 33 | 3,892 | 5,957 | 6 | 28 | 128,521 | 30.0 | 6,032 |
| Omaha | 89.0 | 389,881 | 25.7 | 34 | 4,381 | 6,335 | 23 | 35 | 301,598 | 20.1 | 6,315 |
| *Nevada* | | | | | | | | | | | |
| Las Vegas | 34.3 | 89,427 | n.a. | n.a. | 2,607 | 7,294 | 28 | 36 | 64,405 | 161.6 | 7,662 |
| Reno | 16.3 | 70,189 | n.a. | n.a. | 4,306 | 7,388 | 27 | n.a. | 51,470 | 58.4 | 7,438 |
| *New Hampshire* | | | | | | | | | | | |
| Nashua | n.a. | n.a. | n.a. | n.a. | n.a. | n.a. | n.a. | n.a. | 39,096 | 12.8 | 6,108 |
| Manchester | 34.6 | 91,698 | 8.0 | 1 | 2,650 | 5,802 | 4 | n.a. | 88,282 | 6.7 | 5,796 |

| | | | | | | | | | | | |
|---|---|---|---|---|---|---|---|---|---|---|---|
| *New Jersey* | | | | | | | | | | | |
| Atlantic City | 60.0 | 124,902 | 18.9 | 9 | 2,082 | 5,172 | 52 | 26 | 59,544 | −3.4 | 4,108 |
| Camden | n.a. | n.a. | n.a. | n.a. | n.a. | n.a. | n.a. | n.a. | 117,159 | −5.9 | 5,471 |
| Union | n.a. | n.a. | n.a. | n.a. | n.a. | n.a. | n.a. | n.a. | 52,180 | −6.0 | 5,815 |
| Edison | n.a. | n.a. | n.a. | n.a. | n.a. | n.a. | n.a. | n.a. | 44,799 | 174.0 | n.a. |
| Trenton | 75.3 | 242,401 | 28.0 | 190 | 3,219 | 6,617 | 53 | 6 | 114,167 | −10.8 | 5,840 |
| *New Mexico* | | | | | | | | | | | |
| Albuquerque | 76.6 | 241,216 | n.a. | n.a. | 3,149 | 6,350 | 17 | 55 | 201,189 | 107.8 | 6,621 |
| Las Cruces | n.a. | n.a. | n.a. | n.a. | n.a. | n.a. | n.a. | n.a. | 29,367 | 138.3 | 5,789 |
| Roswell | n.a. | n.a. | n.a. | n.a. | n.a. | n.a. | n.a. | n.a. | 39,593 | 53.8 | 5,543 |
| Santa Fe | n.a. | n.a. | n.a. | n.a. | n.a. | n.a. | n.a. | n.a. | 33,394 | 19.3 | 5,502 |
| *New York* | | | | | | | | | | | |
| Albany | 106.4 | 455,447 | 9.7 | 95 | 4,281 | 6,183 | 72 | 41 | 129,726 | −3.9 | 5,778 |
| Buffalo | 160.2 | 1,054,370 | 17.7 | 59 | 6,582 | 6,394 | 49 | 36 | 532,759 | −8.2 | 5,713 |
| Elmira | n.a. | n.a. | n.a. | n.a. | n.a. | n.a. | n.a. | n.a. | 46,517 | −6.4 | 5,452 |
| Hempstead | n.a. | n.a. | n.a. | n.a. | n.a. | n.a. | n.a. | n.a. | 34,641 | 18.9 | 7,455 |
| Poughkeepsie | 113.3 | n.a. | n.a. | n.a. | n.a. | n.a. | n.a. | n.a. | 38,330 | −6.6 | 5,893 |
| Rochester | n.a. | 493,402 | 20.6 | 75 | 4,355 | 7,098 | 35 | 75 | 318,611 | −4.2 | 6,361 |
| Rome | n.a. | n.a. | n.a. | n.a. | n.a. | n.a. | n.a. | n.a. | 51,646 | 23.9 | 6,255 |
| Schenectady | 106.4 | 455,447 | 9.7 | 95 | 4,281 | 6,183 | 82 | 41 | 81,682 | −11.0 | 5,925 |
| Syracuse | 67.7 | 333,286 | 25.6 | 55 | 4,923 | 6,737 | 35 | 34 | 216,063 | −2.1 | 6,247 |
| *North Carolina* | | | | | | | | | | | |
| Charlotte | 73.9 | 209,551 | 48.7 | 114 | 2,836 | 5,663 | 4 | 60 | 201,564 | 50.4 | 5,592 |
| Durham | 27.0 | 84,642 | 15.4 | 82 | 3,135 | 4,755 | 7 | 68 | 78,302 | 9.8 | 4,673 |
| Fayetteville | n.a. | n.a. | n.a. | n.a. | n.a. | n.a. | n.a. | n.a. | 47,106 | 35.7 | 3,960 |
| Greensboro | 50.8 | 123,334 | 47.9 | 112 | 2,428 | 5,824 | 3 | 60 | 119,574 | 60.7 | 5,845 |
| Raleigh | 33.5 | 93,931 | 36.6 | 172 | 2,804 | 5,586 | n.a. | 39 | 93,931 | 43.0 | 5,586 |
| Rocky Mount | n.a. | n.a. | n.a. | n.a. | n.a. | n.a. | n.a. | n.a. | 32,147 | 16.1 | 4,464 |
| Winston-Salem | 43.0 | 128,176 | 38.6 | 86 | 2,981 | 5,480 | 13 | 70 | 111,135 | 26.6 | 5,317 |
| *North Dakota* | | | | | | | | | | | |
| Fargo | 20.2 | 72,730 | n.a. | n.a. | 3,600 | 6,363 | 36 | n.a. | 46,662 | 22.0 | 6,522 |
| *Ohio* | | | | | | | | | | | |
| Akron | 141.3 | 458,253 | 24.9 | 44 | 3,243 | 6,671 | 37 | 46 | 290,351 | 5.7 | 6,466 |
| Canton | 50.7 | 213,574 | 22.8 | 53 | 4,213 | 6,209 | 47 | 30 | 113,631 | −2.8 | 5,736 |
| Cincinnati | 242.3 | 993,568 | 22.2 | 66 | 4,101 | 6,317 | 49 | 42 | 502,550 | −0.3 | 5,701 |
| Cleveland | 586.7 | 1,784,991 | 29.0 | 96 | 3,042 | 6,967 | 51 | 50 | 876,050 | −4.2 | 5,935 |
| Columbus | 144.8 | 616,748 | 40.9 | 124 | 4,259 | 6,403 | 24 | 48 | 471,316 | 25.4 | 5,982 |
| Dayton | 124.5 | 501,664 | 44.6 | 99 | 4,029 | 6,833 | 48 | 38 | 262,332 | 7.6 | 6,266 |
| Elyria | n.a. | n.a. | n.a. | n.a. | n.a. | n.a. | n.a. | n.a. | 43,782 | 44.5 | 6,486 |
| Lancaster | 13.1 | n.a. | n.a. | n.a. | n.a. | n.a. | n.a. | n.a. | 29,916 | 23.7 | 5,873 |
| Lima | n.a. | · 62,963 | n.a. | n.a. | 4,806 | 5,729 | 19 | 38 | 51,037 | 1.6 | 5,637 |
| Mansfield | n.a. | n.a. | n.a. | n.a. | n.a. | n.a. | n.a. | n.a. | 47,325 | 8.6 | 6,492 |

Appendix Table A–9.  (Continued)

| State and city | Land area (sq. mi.) (7)ᵃ | Urbanized area Pop. (8) | Urbanized area Δ Pop. 1950–60 (%) (3) | Urbanized area Δ Land area, 1950–60 (%) (4) | Urbanized area Pop./ sq. mi. (9) | Urbanized area Income ($) (5) | Urbanized area Fringe pop. (%) (6) | SMSA Pop. in ring, work in city (%) (13) | City Pop. (12) | City Δ Pop., 1950–60 (%) (11) | City Income ($) (10) |
|---|---|---|---|---|---|---|---|---|---|---|---|
| **Ohio (Continued)** | | | | | | | | | | | |
| Sandusky | n.a. | n.a. | n.a. | n.a. | n.a. | n.a. | n.a. | n.a. | 31,989 | 8.9 | 6,064 |
| Springfield | 20.6 | 90,157 | 9.6 | 56 | 4,377 | 5,705 | 8 | 44 | 82,723 | 5.4 | 5,673 |
| Toledo | 134.9 | 438,238 | 20.3 | 93 | 3,249 | 6,579 | 27 | 52 | 318,003 | 4.7 | 6,299 |
| Youngstown | 108.0 | 372,748 | 25.1 | 37 | 3,451 | 6,194 | 55 | 41 | 166,689 | −1.0 | 5,749 |
| **Oklahoma** | | | | | | | | | | | |
| Bartlesville | n.a. | n.a. | n.a. | n.a. | n.a. | n.a. | n.a. | n.a. | 27,893 | 45.1 | 6,606 |
| Enid | n.a. | n.a. | n.a. | n.a. | n.a. | n.a. | n.a. | n.a. | 38,859 | 7.9 | 4,956 |
| Lawton | 13.2 | 61,941 | n.a. | n.a. | 4,693 | 4,631 | n.a. | n.a. | 61,697 | 77.5 | 4,633 |
| Norman | n.a. | n.a. | n.a. | n.a. | n.a. | n.a. | n.a. | n.a. | 33,412 | 23.7 | 5,259 |
| Oklahoma City | 385.2 | 429,188 | 56.0 | 475 | 1,114 | 5,740 | 24 | 46 | 324,253 | 33.2 | 5,600 |
| Tulsa | 70.2 | 298,922 | 44.9 | 86 | 4,258 | 6,175 | 12 | 34 | 261,685 | 43.2 | 6,229 |
| **Oregon** | | | | | | | | | | | |
| Eugene | 38.2 | 95,868 | n.a. | n.a. | 2,510 | 6,231 | 47 | 31 | 50,977 | 42.1 | 6,267 |
| Portland | 192.4 | 651,685 | 27.1 | 70 | 3,387 | 6,522 | 43 | 39 | 372,676 | −0.3 | 6,335 |
| Salem | n.a. | n.a. | n.a. | n.a. | n.a. | n.a. | n.a. | n.a. | 49,142 | 13.9 | 5,859 |
| **Pennsylvania** | | | | | | | | | | | |
| Allentown | 60.1 | 256,016 | 13.3 | 22 | 4,260 | 6,026 | 58 | 34 | 108,347 | 1.5 | 6,049 |
| Altoona | 18.0 | 83,058 | −4.1 | 28 | 4,614 | 5,181 | 16 | 35 | 69,407 | −10.1 | 5,138 |
| Erie | 56.7 | 177,433 | 17.0 | 90 | 3,129 | 5,894 | 22 | 19 | 138,440 | 5.8 | 5,733 |
| Harrisburg | 48.2 | 209,501 | 23.5 | 65 | 4,346 | 6,119 | 62 | 29 | 79,697 | −11.0 | 5,403 |
| Johnstown | 21.0 | 96,474 | 3.3 | 43 | 4,594 | 5,157 | 44 | 21 | 53,949 | −14.7 | 4,674 |
| Lancaster | 29.2 | 93,855 | 23.0 | 256 | 3,214 | 6,272 | 35 | 22 | 61,055 | −4.3 | 5,645 |
| Philadelphia | 596.7 | 3,635,228 | 24.4 | 91 | 6,092 | 6,437 | 45 | 23 | 2,002,512 | −3.3 | 5,782 |
| Pittsburgh | 525.0 | 1,804,400 | 17.7 | 107 | 3,437 | 6,106 | 67 | 16 | 604,332 | −10.7 | 5,605 |
| Reading | 33.1 | 160,297 | 3.5 | 25 | 4,843 | 5,933 | 39 | 26 | 98,177 | −10.2 | 5,453 |
| Williamsport | n.a. | n.a. | n.a. | n.a. | n.a. | n.a. | n.a. | n.a. | 41,967 | −6.8 | 5,228 |
| York | 28.1 | 100,872 | 28.0 | 209 | 3,590 | 5,933 | 46 | 25 | 54,504 | −9.1 | 5,441 |
| **Rhode Island** | | | | | | | | | | | |
| Providence | 188.0 | 659,542 | 13.1 | 32 | 3,508 | 5,688 | 69 | 29 | 207,498 | −16.6 | 5,069 |

| | | | | | | | | | | | |
|---|---|---|---|---|---|---|---|---|---|---|---|
| ***South Carolina*** | | | | | | | | | | | |
| Charleston | 30.8 | 160,113 | 33.1 | 67 | 5,198 | 4,692 | 59 | 24 | 65,925 | −6.1 | 3,708 |
| Columbia | 52.3 | 162,601 | 34.6 | 82 | 3,109 | 4,858 | 40 | 30 | 97,433 | 12.1 | 4,574 |
| Greenville | 52.6 | 126,887 | n.a. | n.a. | 2,412 | 4,869 | 48 | 26 | 66,188 | 13.8 | 4,754 |
| Spartanburg | n.a. | n.a. | n.a. | n.a. | n.a. | n.a. | n.a. | n.a. | 44,352 | 20.5 | 4,483 |
| ***South Dakota*** | | | | | | | | | | | |
| Rapid City | n.a. | n.a. | n.a. | n.a. | n.a. | n.a. | n.a. | n.a. | 42,399 | 67.5 | 5,694 |
| Sioux Falls | 17.4 | 66,582 | n.a. | n.a. | 3,827 | 6,072 | 2 | n.a. | 65,466 | 24.2 | 6,081 |
| ***Tennessee*** | | | | | | | | | | | |
| Chattanooga | 89.1 | 205,143 | 22.3 | 77 | 2,302 | 5,168 | 37 | 51 | 130,009 | −0.8 | 4,438 |
| Jackson | n.a. | n.a. | n.a. | n.a. | n.a. | n.a. | n.a. | n.a. | 34,376 | 13.8 | 4,017 |
| Johnson City | n.a. | n.a. | n.a. | n.a. | n.a. | n.a. | n.a. | n.a. | 31,187 | 11.9 | 4,476 |
| Knoxville | 59.7 | 172,734 | 16.6 | 68 | 2,893 | 4,962 | 35 | 35 | 111,827 | −10.4 | 4,244 |
| Memphis | 155.7 | 544,505 | 34.1 | 42 | 3,497 | 4,991 | 9 | 58 | 497,524 | 25.6 | 4,915 |
| Nashville | 129.3 | 346,729 | 33.9 | 141 | 2,682 | 5,272 | 51 | 51 | 170,874 | −2.0 | 3,816 |
| Oak Ridge | n.a. | n.a. | n.a. | n.a. | n.a. | n.a. | n.a. | n.a. | 27,169 | n.a. | 7,566 |
| ***Texas*** | | | | | | | | | | | |
| Abilene | 63.8 | 91,566 | n.a. | n.a. | 1,435 | 5,438 | 1 | 18 | 90,368 | 98.3 | 5,460 |
| Amarillo | 54.8 | 137,969 | 85.3 | 149 | 2,518 | 5,877 | n.a. | 23 | 137,969 | 85.8 | 5,877 |
| Austin | 50.7 | 187,157 | 37.6 | 47 | 3,691 | 5,116 | n.a. | 47 | 186,545 | 40.8 | 5,119 |
| Beaumont | 73.3 | 119,178 | 26.6 | 99 | 1,626 | 5,577 | 5 | 20 | 119,175 | 26.8 | 5,577 |
| Corpus Christi | 53.1 | 177,380 | 44.3 | 81 | 3,340 | 5,166 | 27 | 35 | 167,690 | 54.9 | 5,221 |
| Dallas | 647.0 | 932,394 | 73.0 | 353 | 1,441 | 6,227 | n.a. | 15 | 679,674 | 56.4 | 5,976 |
| El Paso | 115.0 | 277,128 | 102.4 | 320 | 2,410 | 5,208 | 29 | 35 | 276,687 | 112.0 | 5,211 |
| Fort Worth | 272.6 | 502,682 | 59.3 | 129 | 1,844 | 5,742 | 18 | 43 | 356,268 | 27.8 | 5,484 |
| Houston | 430.5 | 1,139,678 | 62.7 | 59 | 2,647 | 6,054 | n.a. | n.a. | 938,219 | 57.4 | 5,902 |
| Longview | n.a. | n.a. | n.a. | n.a. | n.a. | n.a. | 1 | 18 | 40,050 | 63.5 | 5,355 |
| Lubbock | 76.2 | 129,289 | n.a. | n.a. | 1,697 | 5,575 | n.a. | n.a. | 128,691 | 79.4 | 5,582 |
| Midland | 23.5 | 63,274 | n.a. | n.a. | 2,693 | 7,069 | n.a. | n.a. | 62,625 | 188.4 | 7,094 |
| San Angelo | 29.7 | 58,815 | n.a. | n.a. | 1,980 | 4,650 | 8 | 28 | 58,815 | 12.9 | 4,650 |
| San Antonio | 194.4 | 641,965 | 42.8 | 117 | 3,302 | 4,789 | 1 | n.a. | 587,718 | 43.9 | 4,691 |
| Texarkana | n.a. | n.a. | n.a. | n.a. | n.a. | n.a. | n.a. | n.a. | 30,218 | 22.1 | 4,353 |
| Tyler | 18.6 | 51,739 | n.a. | n.a. | 2,782 | 5,444 | 1 | n.a. | 51,230 | 31.5 | 5,478 |
| Victoria | n.a. | n.a. | n.a. | n.a. | n.a. | n.a. | n.a. | n.a. | 33,047 | 104.9 | 5,279 |
| Waco | 64.9 | 116,163 | 25.1 | 116 | 1,790 | 4,925 | 16 | 41 | 97,808 | 15.5 | 4,859 |
| Wichita Falls | 37.4 | 102,104 | n.a. | n.a. | 2,730 | 5,442 | n.a. | 31 | 101,724 | 49.5 | 5,451 |
| ***Utah*** | | | | | | | | | | | |
| Ogden | 66.7 | 121,927 | n.a. | n.a. | 1,828 | 6,358 | 42 | 36 | 70,197 | 22.9 | 6,145 |
| Salt Lake City | 131.7 | 348,661 | 53.3 | 73 | 2,647 | 6,407 | 46 | 50 | 189,454 | 4.0 | 6,135 |
| ***Vermont*** | | | | | | | | | | | |
| Burlington | n.a. | n.a. | n.a. | n.a. | n.a. | n.a. | n.a. | n.a. | 35,531 | 7.2 | 5,487 |

Appendix Table A–9. (*Continued*)

| State and city | Urbanized area | | | | | | | SMSA | City | | |
|---|---|---|---|---|---|---|---|---|---|---|---|
| | Land area (sq. mi.) | Pop. | Δ Pop. 1950–60 (%) | Δ Land area, 1950–60 (%) | Pop./sq. mi. | Income ($) | Fringe pop. (%) | Pop. in ring, work in city (%) | Pop. | Δ Pop., 1950–60 (%) | Income ($) |
| | (7)[a] | (8) | (3) | (4) | (9) | (5) | (6) | (13) | (12) | (11) | (10) |
| *Virginia* | | | | | | | | | | | |
| Charlottesville | n.a. | n.a. | n.a. | n.a. | n.a. | n.a. | n.a. | n.a. | 29,427 | 13.3 | 5,192 |
| Danville | n.a. | n.a. | n.a. | n.a. | n.a. | n.a. | n.a. | n.a. | 46,577 | 32.8 | 4,883 |
| Lynchburg | 27.6 | 59,319 | n.a. | n.a. | 2,149 | 5,483 | 8 | n.a. | 54,790 | 14.8 | 5,472 |
| Newport News | 149.1 | 208,874 | n.a. | n.a. | 1,401 | 5,771 | 46 | 37 | 113,662 | 168.3 | 5,619 |
| Norfolk | 108.6 | 507,825 | 31.9 | 74 | 4,676 | 5,075 | 40 | 42 | 304,869 | 42.8 | 4,894 |
| Petersburg | n.a. | n.a. | n.a. | n.a. | n.a. | n.a. | n.a. | n.a. | 36,750 | 4.8 | 4,406 |
| Richmond | 88.5 | 333,483 | 29.2 | 83 | 3,768 | 6,037 | 34 | 59 | 219,958 | −4.5 | 5,156 |
| Roanoke | 40.4 | 124,752 | 16.9 | 16 | 3,088 | 5,210 | 22 | 24 | 97,110 | 5.6 | 5,103 |
| *Washington* | | | | | | | | | | | |
| Seattle | 283.3 | 864,109 | 39.0 | 131 | 3,050 | 7,153 | 36 | 39 | 557,087 | 19.1 | 6,942 |
| Spokane | 64.1 | 226,938 | 28.9 | 25 | 3,540 | 6,211 | 20 | 39 | 181,608 | 12.3 | 6,044 |
| Tacoma | 82.8 | 214,930 | 28.2 | 33 | 2,596 | 6,077 | 31 | 45 | 147,979 | 3.0 | 5,993 |
| Yakima | n.a. | n.a. | n.a. | n.a. | n.a. | n.a. | n.a. | n.a. | 43,284 | 12.5 | 5,900 |
| *West Virginia* | | | | | | | | | | | |
| Charleston | 59.9 | 169,500 | 29.5 | 106 | 2,830 | 6,342 | 49 | 32 | 85,796 | 16.7 | 6,146 |
| Huntington | 42.2 | 165,732 | 6.0 | 14 | 3,927 | 5,321 | 50 | 36 | 83,627 | −3.2 | 5,426 |
| Wheeling | 27.3 | 98,951 | −7.2 | 18 | 3,625 | 5,304 | 46 | 16 | 53,400 | −9.3 | 5,392 |

| | | | | | | | | | |
|---|---|---|---|---|---|---|---|---|---|
| *Wisconsin* | | | | | | | | | |
| Appleton | n.a. | n.a. | n.a. | n.a. | n.a. | n.a. | n.a. | 42.3 | 6,515 |
| Beloit | n.a. | n.a. | n.a. | n.a. | n.a. | n.a. | n.a. | 11.0 | 6,063 |
| Green Bay | 46.6 | 97,162 | n.a. | 2,085 | 6,162 | 35 | 41 | 19.3 | 5,981 |
| Madison | 54.3 | 157,814 | 122 | 2,906 | 6,928 | 20 | 42 | 31.9 | 6,799 |
| Milwaukee | 392.0 | 1,149,997 | 285 | 2,934 | 7,036 | 36 | 46 | 16.3 | 6,664 |
| Racine | 14.6 | 95,862 | 21 | 6,566 | 6,780 | 7 | n.a. | 25.2 | 6,758 |
| *Wyoming* | | | | | | | | | |
| Cheyenne | n.a. | n.a. | n.a. | n.a. | n.a. | n.a. | n.a. | 36.2 | 6,575 |
| Summary of data used in simple correlations in Table 17 | | | | | | | | | |
| Mean | 128 | 454,176 | 94 | 3,349 | 5,935 | 31 | 38 | 29 | 5,667 |
| Standard deviation | 183 | 835,062 | 78 | 1,075 | 725 | 21 | 12 | 53 | 819 |
| No. of observations | 171 | 171 | 134 | 171 | 171 | 168 | 153 | 258 | 256 |
| Summary of data used in multiple correlations in Table 18 | | | | | | | | | |
| Mean | 152 | 546,388 | 95 | 3,412 | 5,934 | 34 | 39 | 18.7 | 5,633 |
| Standard deviation | 198 | 917,071 | 78 | 1,029 | 699 | 20 | 12 | 36.8 | 715 |

n.a. = Not applicable.

ᵃ Column numbers correspond with those in Tables 17 and 18 in the text.

*Sources* (identified by column numbers):

3. U.S. Bureau of the Census, *County and City Data Book, 1962* (Washington: U.S. Government Printing Office, 1962). Table 4, Column 205.

4. *Ibid.*, computed from variable 5 below and 1950 data from *U.S. Census of Population, 1950*, Vol. 1, *Number of Inhabitants*, Ch. 1, *U.S. Summary* (Washington: U.S. Government Printing Office, 1952), Table 17, pp. 26–29.

5. *Ibid.*, Table 4, Column 218.

6. U.S. Bureau of the Census, *U.S. Census of Population, 1960, U.S. Summary* (Washington: U.S. Government Printing Office, 1961), *General Social and Economic Characteristics*, PC(1)–1C. Computed from Table 151, Column 1.

7. U.S. Bureau of the Census, *County and City Data Book, 1962, op. cit.*, Table 4, Column 201.

8. *Ibid.*, Table 4, Column 203.

9. *Ibid.*, Table 4, Column 204.

10. *Ibid.*, Table 6, Column 320.

11. *Ibid.*, Table 6, Column 305.

12. *Ibid.*, Table 6, Column 303.

13. U.S. Bureau of the Census, *U.S. Census of Population, 1960, U.S. Summary* (Washington: U.S. Government Printing Office, 1961); *Detailed Characteristics*, PC(1)–1D, Table 302, Column 11.

For Product Safety Concerns and Information please contact our EU
representative GPSR@taylorandfrancis.com Taylor & Francis Verlag GmbH,
Kaufingerstraße 24, 80331 München, Germany


Printed and bound by CPI Group (UK) Ltd, Croydon, CR0 4YY

13/05/2025

01870224-0001